Spaces for children
儿童空间

建筑立场系列丛书 No.92

[美] 玛莎·索恩 等 | 编
司炳月 凌玥瑶 王晓华 | 译

大连理工大学出版社

儿童空间

C3 建筑立场系列丛书 No.92

004	教育空间与场所 _ Martha Thorne
012	Five Fields游乐设施 _ Matter Design + FR|SCH Projects
020	因斯布鲁克大学的游戏屋 _ Studio3-Institute for Experimental Studies UIBK
034	格利法达公立幼儿园 _ KLab Architecture
046	瓜斯塔拉幼儿园 _ Mario Cucinella Architects
054	NUBO _ PAL Design Group
064	Niederolang幼儿园 _ Feld72
076	ATM幼儿园 _ Hibinosekkei + Youji no Shiro
086	Chuon Chuon Kim第二幼儿园 _ KIENTRUC O
100	Fukumasu基地 _ Yasutaka Yoshimura Architects
112	童戏空间 _ Alison Killing
120	朝日幼儿园 _ Tezuka Architects
130	希望学校 _ Grupo Garoa Arquitetos Associados
144	巴兰基亚幼儿园 _ El Equipo Mazzanti
158	埃克苏佩里国际学校 _ 8 A.M.
174	玛利娅·蒙特梭利马萨特兰学校 _ EPArquitectos + Estudio Macías Peredo
188	儿童村 _ Aleph Zero + Rosenbaum
206	be-MINE游乐场 _ OMGEVING + Carve
220	建筑师索引

Spaces for Children

C3 No. 92 Spaces for Children

004   Spaces and Places for Education _ Martha Thorne

012   Five Fields Play Structure _ Matter Design + FR|SCH Projects

020   Playroom in University of Innsbruck _ Studio3 - Institute for Experimental Studies UIBK

034   Public Nursery in Glyfada _ KLab Architecture

046   Nursery in Guastalla _ Mario Cucinella Architects

054   NUBO _ PAL Design Group

064   Kindergarten Niederolang _ Feld72

076   ATM Nursery _ Hibinosekkei + Youji no Shiro

086   Chuon Chuon Kim 2 Kindergarten _ KIENTRUC O

100   Fukumasu Base _ Yasutaka Yoshimura Architects

112   Child's Play _ Alison Killing

120   Asahi Kindergarten _ Tezuka Architects

130   Wish School _ Grupo Garoa Arquitetos Associados

144   Baby Gym Barranquilla _ El Equipo Mazzanti

158   Exupéry International School _ 8 A.M.

174   María Montessori Mazatlán School _ EPArquitectos + Estudio Macías Peredo

188   Children's Village _ Aleph Zero + Rosenbaum

206   Play Landscape be-MINE _ OMGEVING + Carve

220   Index

Space Child

儿童空间

教育空间与场所_Spaces and Places for Education / Martha Thorne
童戏空间_Child's Play / Alison Killing

教育空间与场所
Spaces and Places for Education

Martha Thorne

> 根本不存在所谓的"中性"环境：你身边的建成环境要么成就你，要么伤害你。　　——Sarah Goldhagen

显而易见，儿童教育空间的发展在这些年来已经取得了进步，但是这条路并非一帆风顺。教育领域、心理学领域、设计领域和政治领域都对此提出了不同的理论和观点。然而，如今为了了解教和学的过程及实体环境的意义，我们在研究中一直不断地探索并分享信息，这是以往任何时候都不曾做到的事情。当我们逐渐意识到实体环境远比我们想象中的重要时，我们也许正在见证一场思想模式的转变。

回顾历史可知，19世纪教师弗里德里希·福禄贝尔（Friedrich Fröbel）就是那些首批认识到身体活动在儿童教育过程中扮演重要地位的人之一。他提出，娱乐是学习的一部分。他还发明了多种教育玩具，其中就有（广为人知的）福禄贝尔积木。

玛利娅·蒙特梭利（Maria Montessori）或许是与儿童发展和创新教育相关的最著名人物。她是一名内科医生，也是一名教育家。她的教育方法在20世纪早期就得到了应用，除了她的祖国意大利，还有许多其他地方，甚至是欧洲以外的一些地区都对其理论表示认同。她的教育方法以儿童为中心，并以广泛的原则为基础，即：儿童会在与环境的互动过程中完成心理上的自我建构，与此同时，儿童和青少年与生俱来地拥有自发学习的内在要求。

There's no such thing as a "neutral" environment: your built environment is either helping you or hurting you.
– Sarah Goldhagen

Clearly educational spaces for children have evolved over the years. However, the path has neither been constant, nor smooth. Different theories and ideas have surfaced in different fields – education, psychology, design, and politics. However, today as in no time in history, are we developing and sharing information in our search to understand the processes involved in learning and teaching and the role of the physical environment. We may be witnessing a paradigm shift as we realize that the physical settings play a much larger role than ever considered.

When glancing back at history, the name 19th-century pedagogue Friedrich Fröbel appears as one of the first who recognized the importance of physical activity in the learning process of young children. He proposed the idea of play as part of the learning process and invented, among other educational toys, the Frobel blocks.

Perhaps the most well-known name associated with children and innovative educational advancements is Maria Montessori, a physician, and educator whose method was adopted in the early 20th century and took hold in many different places outside her home country Italy and even beyond Europe. Her method was child-centered and was based on broad principles: children engage in psychological

荷兰阿姆斯特丹的蒙特梭利学校，1996年，赫曼·赫茨伯格
Montessori School, Amsterdam, The Netherlands, 1966, Herman Hertzberger

美国伊利诺伊州温尼卡镇的乌鸦岛学校，1940年，Perkins Wheeler & Will与埃罗·沙里宁联合设计
Crow Island School in Winnetka, Illinois, USA, 1940, Perkins, Wheeler & Will with Eero Saarinen

  蒙特梭利认为，儿童在有备环境中学习，并为自己做事，效果才能最好。有备环境使孩子们可以有序获得学习材料和学习经验。蒙特梭利所描绘的教室，也正是众多教育工作者在谈到以儿童为中心的教育和主动学习的问题时所提倡的。此种环境最核心的特点就是自由。在这种环境中，孩子们可以自主选择材料并进行探究，进而吸收他们所发现的内容和知识。玛利娅·蒙特梭利非常擅长为儿童们创造这种环境，使他们成长为积极、独立并热爱学习的人。¹

  渐进式教育在20世纪取得了长足的进步。世界各地不同国家的很多学者都曾尝试改善教学方法，提升教学理解。渐进式教育与传统的教学方法不同，它强调解决问题的能力，强调合作，鼓励孩子们参加课堂内外各种多变的活动，强调对话与娱乐。它还相信教育是民主的一部分，不应该一再重复现有的课程和社会结构。此外，它还提出了许多其他观点。它在20世纪80年代所提出的一种观点认为，孩子们应积极参与学习过程，包括实验室内进行的科学课以及语言课，这一观点与"实践出真知"不谋而合。

  在20世纪，有两个因给儿童设计的学习空间具有创新性和周全性而闻名的名字，分别是荷兰建筑师赫曼·赫茨伯格（Herman Hertzberger）和美国的珀金斯&威尔公司（Perkins and Will，前身为珀金斯、惠勒&威尔公司，即Perkins Wheeler & Will）。赫茨伯格建造了多所蒙氏学校，但其实他所有的建筑，而不仅仅是教育设施，都体现了灵活性和适应性的设计哲学。他将空间理解为一种能让居住者自行进行解读和定义的框架。

self-construction through interaction with the environment, and children and young people will innately seek to learn spontaneously.

Montessori believed that children learn best in a prepared environment, a place in which children can do things for themselves. The prepared environment makes learning materials and experiences available to children in an orderly format. Classrooms Montessori described are really what educators advocate when they talk about child-centered education and active learning. Freedom is the essential characteristic of the prepared environment. Since children within the environment are free to explore materials of their own choice, they absorb what they find there. Maria Montessori was a master at creating environments for young children that enabled them to be independent, active, and to learn.¹

The 20th century saw further development of ideas of progressive education. There were attempts by many people in numerous countries around the world to advance pedagogical methods and understanding. Progressive education, in contrast to traditional methods, placed emphasis on problem-solving, collaboration, varied types of activities inside and outside the classroom, conversation and play, and a belief in education that as part of democracy, it should not just replicate the existing class or societal structure, among other ideas. In the 1980s, the idea of students actively participating in learning, with the inclusion of laboratories for science classes or languages, showed that "learning by doing" was a true step in the right direction.

在美国,Perkins Wheeler & Will公司则和埃罗·沙里宁(Eero Saarinen)于1941年联合设计了乌鸦岛学校,该学校位于芝加哥北部郊区。此项设计的灵感来源于对教师的访谈和对学生们的观察结果。设计为每间教室都配备了各自的户外庭院,同时学校的每座建筑都有自己的操场。

这并不意味着所有以儿童或学生为设计中心的项目都具有创新性。批判的观点总是如影随形,足以限制许多项目的野心,让他们不得不在只使用一些小尺度的家具和明亮的颜色以后就作罢了。还有一些建筑师和客户则一直秉持一些错误观点,认为只要使用与孩子们身高相匹配的家具或者创造一个"儿童友好型"的环境,就足以声称自己设计的是创新型教室,也足以反映出自己对儿童使用者们的关注和关心了。

不过令人欣慰的是,近年来,建立在真实调查基础之上的高品质学习空间设计正东山再起。建筑师和建筑学也从临时抱佛脚般的设计尝试和看重建成视觉效果转变到以更为复杂、深入的方法来对待学习空间的设计。我认为,这种演进与科技的进步、"用户体验"这一广为传播使用(尤其是在商业领域)的术语、建筑学专业对更多跨学科信息的输入所持的开放态度以及心理学和神经系统科学领域方面的认真研究不无关系。

科技不仅能帮助我们设计出配备更多教与学工具的教室,还陪伴着今天的年轻人一路成长,并使这些年轻人习惯于独立探索、交流与追求快乐生活。与我们的父辈相比,他们所能体验到的刺激范围更广,速度更快。很明显,人们从很小的时候就已经开始运用科技了。罗伯特·贝希纳(Robert Beichner)[2]的调查显示,4岁以下会使用电脑的儿童约占75%,8到

Two architectural firms which have been known for their innovative and thoughtful design of learning spaces for children in the 20th century are Dutch architect Herman Hertzberger and the U.S. firm of Perkins and Will (previously Perkins Wheeler & Will). Hertzberger built several Montessori schools, but all of his buildings, not just educational facilities, embraced a philosophy of flexibility and adaptability in which space is understood as a framework that should enable users to interpret and define how they inhabit it.

In the U.S., Perkins, Wheeler & Will with Eero Saarinen designed Crow Island School (1941) in a suburb north of Chicago. The design is a result of interviews with teachers and observation of students. The result allows each classroom to have its own outdoor courtyard and each wing also has its own playground.

This is not to say that all projects that appear to have children or the students as the focal point are innovative. Critical analysis is warranted for many projects which limit their ambitions to just using small-sized furniture and bright colors. Some architects and clients hold on to the erroneous idea that using furniture that is scaled of a child's body or giving the appearance of "kid-friendly" environments is sufficient to claim an innovative classroom and reflect their concern and attention toward the young users.

Thankfully, the design of quality learning spaces based on real research and evidence has experienced a resurgence in recent years. Architects and architecture have progressed from ad hoc attempts and visual outcomes to more sophisticated and deeper approaches to learning spaces. This evolution, I believe, has something to do with the rise in technology, the widespread acceptance and use (especially in business) of the term "user experience", and an opening up of the architecture profession to more

18岁的典型青少年每天要花7.5个小时对着屏幕，平均每5分钟就要发一条信息。他接着说："科技改变了学生们的思维方式。比起回忆起信息本身的内容，人们更善于记住查找信息的方法。从现实层面讲，我们的认知能力在进化。进化是指一个有机物由于所处环境的变化，其本身的某些方面也开始出现变化的现象。在这种情况下，信息无处不在，我们不用记住所有事实，只需要知道怎样找到它们。"

用户体验（缩写为UX）本质上是一种以人或以利益相关者为中心的方法，这种方法在促成对学习空间的设计朝积极方向的转变中曾起到了一定的作用。在产品和服务的设计与销售过程中，用户体验作为一种了解终端用户的方式被经常运用于商业活动中，它试图更加贴近消费者的价值观、需求和追求。公司会致力于为消费者、客户和用户营造出更有意义、更具个性化的体验，并希望能藉此提高销量，或在某些过程中为客户提供更为满意的体验。在建筑领域，客户在情感和情绪上的共鸣才是最重要的驱动因素，而非最终的产品和功能。换句话说，在当今时代，我们对一个空间的感受如何，对其做出怎样的反应，怎样与环境互动，这些已经与简单地定义空间或通过标准化测试来衡量学习目标变得同等重要。

今天的教育机构正面临挑战，需要对正式或者非正式的学习空间进行重新思考。这其中的有些问题来源于建筑师和设计师，其他一些认知则生发于对相关学科的研究。在环境的创造过程中，必须同时考虑学生、教师和教员的身心健康。

教育空间不应仅是包括教室、图书馆、走廊、体育馆、办公室和实验室在内的一系列功能性空间；它还要做到能鼓励人们去探索，拥有较强的灵活性，并通过允许一系列行为和关系的自主生发进而产生意想不到的用途。教育方法和教室

cross-disciplinary inputs along with the serious research being developed in psychology and neurosciences.

Technology not only allows us to design classrooms that have many more tools for teaching and learning; today's youngsters have also grown up with technology and are accustomed to exploration on their own, communication, and instant gratification. The type of stimuli that they experience has a wider range and much quicker speed than that of our parents. It is obvious that people use technology from a very young age. According to Robert Beichner[2], about 75% of children under age four use computers and the typical 8- to 18-year-old spend seven and a half hours per day watching a screen. They average an outgoing text message every five minutes. He goes on to say, "Technology has literally changed the way students think. People are often better at remembering how to find information than recalling the information itself. In a very real sense, our cognition is evolving. Evolution is the changing of some aspect of an organism because of a change in its environment. In this case, information is everywhere so we don't need to remember all the facts. Just remember how to locate them."

User experience (or UX), which basically seeks to keep people or stakeholders at the center of the approach, has contributed to a positive shift in the design of learning spaces. User experience is often used in business as a way to understand the end users when creating and marketing products and services. It is an attempt to get closer to the values, needs, and aspirations of consumers. Companies seek to create a more meaningful and personalized experience for the customer or client or user. This is hoped to result in more sales, and for certain processes, just a more satisfying experience. Feelings and an emotional connection, opposed to just final product or function, in the case of architecture, become the important drivers. In other words, how one feels, how one responds to a space, and how one

的布局应保持同步。成排的座位设计有助于将关注点放在教师身上,并限制学生们的消极行为;圆形或半圆形的座位安排则有助于师生间的互动。如果空间未经结构设计,或将教学法与外部环境完全对立,则会对老师和学生双双造成压力。教育不应再被单纯视为吸收知识的过程,而应被视为促进整个人的身心发展和精神发展的过程。

今天的我们很幸运,因为已有越来越多的教育家、科学家和建筑师在通力合作。大家共同努力推动研究,在创造最优教育环境的过程中竞相迸发出各种新思路。我们现在可以通过不同渠道获取有意义的数据,这有利于解读环境对人的影响。行为心理学和认知心理学领域的合作还可以帮助我们认识环境的设计可对人的行为、思想和感情所能产生的影响。对建筑师和设计师而言,神经系统科学更是具有无上潜力,可帮助他们了解大脑会如何认识周围的环境,以及在面对不同的刺激和情景时,我们的认知和心理会做出何种反应。

对于那些仍然需要成年人照顾的小孩子们来说,好的环境能让他们在生活和学习中有安全感,这对他们来说至关重要。毫无疑问,学习环境必须要安全健康(无论是在建材、自然光还是与外部的联系等方面)。此外,那些可以使用多种教学方法的空间和那些可以支持不同方法和行为的空间才是当今最成功的设计范例。

interacts with the environment are as important today as simply defining spaces or measuring learning goals through standardized tests.

Educational institutions are today challenged to rethink the formal and informal spaces where learning takes place. Some of this questioning comes from architects and designers, while other knowledge is being generated through studies in adjacent disciplines. The physical and mental well-being of students, teachers, and staff must go hand in hand in the creation of the environment.

Educational spaces are not just a series of functional spaces: classrooms, libraries, corridors, gymnasiums, offices, laboratories; it must encourage discovery, promote flexibility and even unexpected uses by allowing a range of behaviors and relationships to develop in a more spontaneous way. Educational methods and classroom layouts need to be in synch. Seats arranged in rows focus the attention on the teacher and limit more passive behavior of students. Circular or semicircular arrangements foster exchanges with the teacher. A totally unstructured space or a stark contrast in pedagogical method and the physical environments causes strain on both the teacher and the students. Education is no longer seen as just the absorption of knowledge, but rather the education of the entire person – mind, body, and spirit.

Today, we are fortunate that there is much more collaboration among educators, scientists, and architects. These combined efforts are advancing research and testing out new ideas for creating the best learning spaces. We are now able to gather meaningful data from a variety of sources to help interpret the environmental effects on occupants. Collaboration between behavioral and cognitive psychology helps to understand the impacts that design can have on the occupants' actions, thoughts, and feelings. Neuroscience holds great promise for architects and designers to uncover how our brains perceive our surrounding environments and understand our cognitive and physiological responses to different stimuli and situations.

For younger children still very dependent on the care and nurturing of adults, environments that foster

学生和教育者们需要感受到他们与空间之间的联系，而这种感觉的来源就是他们要有控制和改变教育空间某些部分的能力。他们同样需要选择权，需要一系列的设施和家具，需要合作型工作、玩耍、沉思、阅读或者是隐私空间。在设计教育空间时，听觉、温度和视觉等因素都应纳入考虑范畴。我们还日益发现人们与自然环境接触的需求以及模糊室内外边界的需求也在增加。

总的来说，随着我们日渐步入终身学习型社会，关于怎样学习、在哪里学习的研究和信息也越来越多，我们的社会关于学习过程和学习环境的设计要求必将改变。不存在所谓的万能型空间或者万能型教学方法，能够满足各种年龄段、各种类型的学生和各种学科的需求。在设计学习空间时，应考虑到学生的生命需求、神经认知需求和更广泛的健康概念，这样才能为使用者提供最好的效果。除了建筑师和设计师外，我们还要进一步扩大合作团队，共同努力确定空间的需求、最终效果和特征，这样我们才能促成学习空间设计方法的积极转变，进而使空间激发学生好奇心和创造力的能力得以进一步提升。

a sense of security for wellbeing and learning are key. It goes without saying that spaces must be safe and healthy (in terms of materials, natural light, connection to the outdoors, etc.) Beyond that, today's most successful spaces are those where a range of teaching methods can be used and the spaces configured to support those methods as well as a variety of behaviors.

Students and educators need to feel connected to their spaces, and this is gained through their ability to control and change certain aspects of the space. They also need choices and therefore, a range of settings and furniture, spaces for collaborative work, play, meditation, reading or privacy. The acoustical, temperature and visual characteristics must always be considered. Increasingly we are seeing the need for connections with the natural environment and softening the boundaries between interior and exterior.

In summary, as we move toward societies where lifelong learning becomes a reality, and with the enormous increase in research and information about how and where learning occurs, our society will require changes in the design of learning processes and learning environments. No longer will "one size fits all" spaces or one teaching method serves the needs of different ages, types of students and different subject matters. Biological needs, neurobiological processes and the broader concept of wellness should be taken into account when designing learning spaces to provide the best results for all users. By enlarging the teams that work together, beyond architecture and design, to define the needs, outcomes, and characters of spaces, we can create a positive shift in the approach to the design of learning spaces to enhance the curiosity and creativity of our students.

1. G.S. Morrison, *Principles of the Montessori Method* (Pearson Allyn Bacon Prentice Hall, Updated on 30 April 2014).
2. Robert J. Beichner, 'History and Evolution of Active Learning Spaces', *New Directions for Teaching and Learning* (2014 Wiley Periodicals, Inc.), Volume 2014, Issue 137, pp.9-16.

Five Fields 游乐设施
Five Fields Play Structure

Matter Design + FR|SCH Projects

　　这座游乐设施坐落于一处公共场所的景观斜坡上，周边儿童都把这里当成可以共享的后院。20世纪50年代早期，Collaborative建筑师事务所（TAC）在马萨诸塞州的莱克星顿设计并开发了Five Fields社区，并希望通过建造一处公共场地的方式来增强社区的凝聚力。在过去的60多年里，社区居民来了又走，但社区对这片公共用地的关爱却从未改变。TAC一直将这次社区设计视为一种实验，而社区里的人们则希望这种实验精神能延续下去，为此他们希望能为孩子们设计一个既安全又不失乐趣的游戏空间。他们想要一个能给孩子们带来挑战的设施，同时还希望设施里不要带有任何功能性构件。

　　游乐设施的目的在于让孩子们进行探索，这样的设计既新颖，同时也颇富挑战性。

　　这个游乐设施可以通过游戏来培养孩子和成人的想象力。也正是因为这个原因，整个空间都是按照儿童的尺寸制作的，但成人也可以使用。将设施的尺寸缩小后可以减缓成人的速度，但小孩子们却可以在其中随意驰骋。

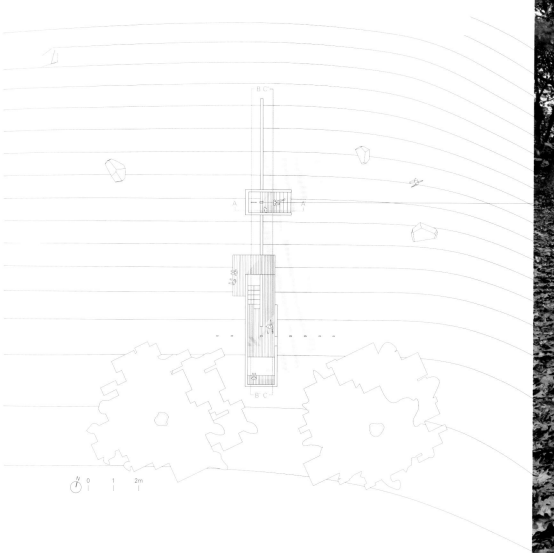

建造此设施的目的就是将使用者从特定的、有目的的使用方式中解脱出来,并希望所有使用者能够在此展现出带有个人色彩的创造性表达方式。使用者可以在游乐装置中发挥想象力,使身体在这一木质空间中得到解放。多样的构件集合在一起,增强了每个使用者的体验。正如每个孩子在学校的学习方式都不同一样,这种游戏环境也为孩子们提供了不同但平等的参与方式。视觉、听觉和动觉元素为使用者创造出独特的瞬间。这里有看不完的景色、做不完的运动、用不完的空间。有些"悬浮"部分相互重叠,形成更为隐蔽的空间。年龄小的孩子可以伏地爬行,年龄较大的孩子则可以向上攀爬,不论是哪种方式,都可以让他们发现可供探索的新空间。设计师用色彩斑斓的图案暗示装置的入口位置,而没有运用明示的方法。门和楼梯等元素也确实存在,但却不是通常想象中的那样。整个设计的意图就是要让建筑变得有童趣一些。在游乐设施中的玩耍在到达死胡同时不得不停下来,但对孩子们来说,这里又可以变成他们眺望远景的好地方。

可以把"Five Fields"的社区游乐设施设计视为一次学习的机会、一次实验。它不仅是对TAC所强调的社区共享理念的认可,也是对它的一种延续。这个空间汇聚了集体的想象力,是一次共享的体验。项目致力于加强社区与公共用地和邻里之间的关系。项目还继承了TAC的传统,对以前的游戏设施(包括秋千、滑梯和沙箱)进行了实验性的延伸。这种对模糊和抽象性的研究不仅仅是一次创造性的尝试,也为项目本身想要体现出的丰富想象力开启了大门。不要问这个设施是用来做什么的,想一想它能激发出什么样的可能性吧。

This play structure is situated on the sloping landscape of a common land, where neighborhood kids enjoy a shared backyard. The Architects Collaborative(TAC) designed and developed the Five Fields neighborhood in Lexington, MA, in the early 1950s, hoping to foster the community by creating a piece of shared land. In the ensuing 60+ years, homeowners have come and gone but the community and its care and appreciation for the common land remain. TAC conceived the neighborhood as an experiment and the community, wishing to keep the experimental spirit alive, requested a structure that is both safe and exhilarating for the kids. They wanted something that would challenge the kids without any singularly functional elements.

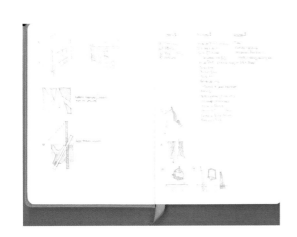

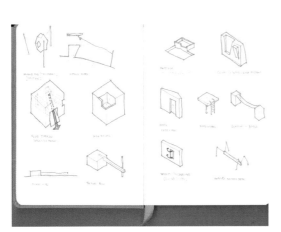

Designing the play structure intended for childish exploration is unique and challenging.

The structure cultivates the child's and the adult's imagination through play. For this reason, these spaces are tailored to children's size but still accessible to adults; it has been shrunk in order to slow the adults, but liberate the kids to fly through the spaces.

The structure was created with the hope to release users from specified, directive uses. The expectation is that all users will inscribe their individual, creative means of expression through the space. The play structure harnesses the imagination and liberates the body within its wooden volumes.

Diverse elements engage the senses and augment each user's experience. Just as each child learns differently in school, this playscape offers diverse but equal means of engagement. Visual, auditory and kinesthetic elements create distinct moments of use. The playscape is an unfolding of views, movements, and spaces. Levitating volumes overlap at certain points to create hidden thresholds. Smaller children crawl, while larger ones climb, which is a varied means of reaching discoverable spaces. Colorful graphics suggest entries without being overtly prescriptive. Elements like doors and stairs exist, but not as expected. The intent was to render architecture childish. Movement through the playscape culminates in dead ends. But to the children, they are vistas of discovery that look out to the landscape.

The Five Fields Play Structure can be seen as a learning opportunity, an experiment. The play structure is both the nod to and continuation of TAC's emphasis on community sharing. This is a space celebrating the collective imagination as a shared experience. It seeks to reinforce the neighbourhood's relationship to the land and to each other. Continuing TAC's heritage, the play structure is an experimental extension of pre-existing play equipment (which already includes swings, slides, and a sandbox). This investigation in ambiguity and abstractness is not just a creative endeavor. It also serves to open the very playful imagination that it is designed for. The intent is not to ask what the structure does, but how it imagines new possibilities.

项目名称：Five Fields Play Structure / 地点：Lexington, Massachusetts, USA / 建筑设计：Brandon Clifford - Matter Design; Michael Schanbacher - FR|SCH Projects / 平面设计：Johanna Lobdell / 项目团队：Dar Adams, Courtney Apgar, Josh Apgar, Michael Leviton, Daniel Marshall, Chris McGuiness, Dan Roseman / 儿童顾问：Liam Apgar, Mack Apgar, Bella Dubrovsky, Sam Leviton, Ainsley Schanbacher, Judson Schanbacher / 尺寸：10.68m² (11m long, 5.5m tall, 1.2m wide, 20m zipline) / 用途：play structure 材料：timber / 竣工时间：2016 / 摄影：courtesy of the Matter Design

A-A' 剖面 section A-A'

B-B' 剖面 section B-B'   C-C' 剖面 section C-C'

因斯布鲁克大学的游戏屋
Playroom in University of Innsbruck
Studio3 – Institute for Experimental Studies UIBK

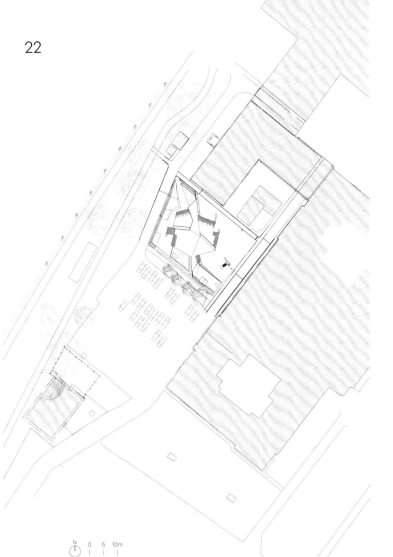

新游戏屋首次高调登台，位置显眼。这里是鼓励发现的地方，是拓展新视野的地方，是孩子们的天堂，全年开放。游戏屋位于因斯布鲁克大学和因河之间，景观多样。学校的员工和学生们设计并建造了这个属于孩子们的新乐巢。

因斯布鲁克大学实验建筑与建筑技术学院第三工作室的21名学生设计建成了这一名为Spielraeume的项目，仅耗时16周，这也是他们学士学位研究项目的一部分。

在沃莱纳·劳赫(Verena Rauch)教授和沃尔特·普莱那(Walter Prenner)教授的指导下，学生们在2016年10至12月间共提供了20个初步设计方案，雅各布·韦斯弗莱克(Jakob Wiesflecker)的设计被确定为最终建造方案。该方案名为"Spielraeume"。在众多公司和专业规划师的支持下，加之学生们的不懈努力，这个凝聚着集体智慧的作品于2017年9月被移交给因斯布鲁克大学和儿童办公室。

设计中的木建筑不禁会让人联想到鸟巢。它的两侧引导墙与周边建筑相呼应，其上带通高的落地窗，露台则面向因河步道开放。游戏屋的围栏被安装在学生们用水泥砌成的矮墙内，用来保护房子那有趣的外部空间免遭侵占，同时还在公共区域和日托中心的隐私之间开辟了一处过渡区域。

一条铺好的小路从项目的南角通向带顶的入口。

室内空间从入口坡道处展开，像三叶草的叶子一样展开到其他房间。空间的连续性设计既允许所有必要空间的离散存在，同时也允许不同区域的空间进行合并。在右手边，越过坡道，经过衣帽间就是办公室，办公室被玻璃墙围了起来。毗邻的家长公共区域由一个插入的开放式基础设施块与会议室分隔开。基础设施块只占房间高度的三分之二，包括几间休息室，一个更衣台，一个厨房和一个杂物间，中间都有

过道。建筑的左手边有两间教室，阳光充足，光线可以从通高的落地窗和天窗照射进来。天窗和落地窗在天花板处相接，反映出空间的连续感。位于河边一侧的户外区域可通过重蚁木露台进入，该露台可以通过坡道或第一间教室到达。

整座建筑完全采用木制结构。为了使建筑成本降到最低，所有的承重组件都选用云杉层压交叉木材（简称BBS）。这种建材也省时，可以让学生们在与温克勒（Winkler）公司的合作下迅速完成此项目。鸟巢状木质板条包裹着具有隔热效果的外墙。建筑内部的木头则全是未经处理的原木。所有的家具都由三层云杉木制成，由学生们自己设计安装。建筑的内部与外部在形式和建材上都保持一致。

Something new stirs for the first time – taking the stage at a prominent location, it is a place for discovery and encouragement, a place for new perspectives, a place for children and a house for all seasons. Situated in a diverse landscape between the University of Innsbruck and the Inn River, this new nest for children was designed and constructed by the university's own staff and students.

It took only 16 weeks for 21 students of studio3, the Institute for Experimental Architecture and Building Technology at the University of Innsbruck, to design and build the Spielraeume as a part of their bachelor's program.

Under the supervision of Professor Verena Rauch and Walter Prenner, the students produced 20 preliminary designs between October and December of 2016, with Jakob Wiesflecker's design being selected for construction. The selected design, called "Spielraeume", was developed collectively and handed over to the University of Innsbruck and the children's office in September of 2017 with the support of numerous companies and specialist planners, and the dedicated efforts of the students themselves.

The wooden structure in design is reminiscent of a nest. Its two guiding walls respond to the surrounding buildings with floor-to-ceiling windows and terraces that open towards the Inn promenade. The enclosure, set in concrete by the students, protects the playful exterior space of the playroom from being overrun by the outside world, while creating an interface between the public sphere and the intimacy of the daycare center.

A paved pathway leads from the southern corner of the property to the covered entrance.

The interior space unfolds from the entrance ramp and expands into the rest of the rooms like the leaves of a clover. The spatial continuum allows the discrete presence of all

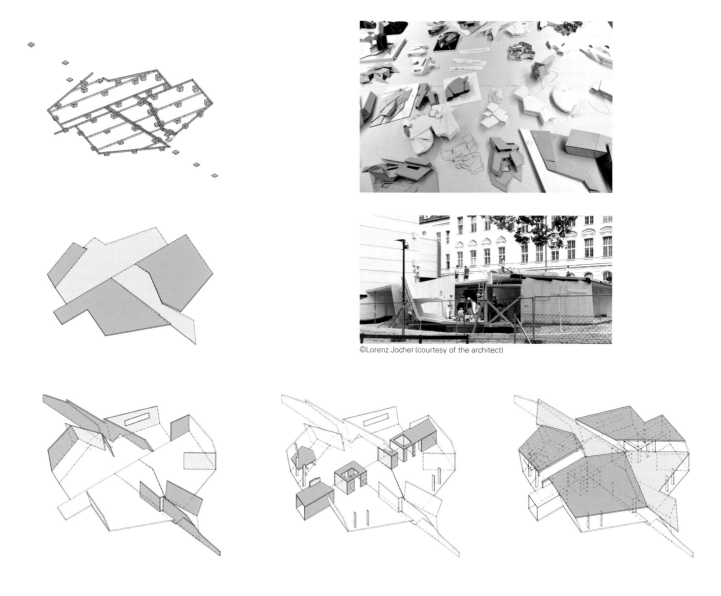

©Lorenz Jocher (courtesy of the architect)

项目名称：Spielraeume - Playrooms / 地点：Innrain 52a, 6020 Innsbruck, Austria / 建筑设计：studio3 - Institute for Experimental Studies UIBK / 学生：Baio Mario, Barbisch Melissa, Brun Fabio, Bugelnig Niklas, Decker Nicolas, Dettler Jonas, Grimm Raffael, Hörl Andreas, Jocher Lorenz, Kegel Alexander, Mayr Thomas, Moschig Verena, Neuwirth Sabrina, Nikisch Janis, Obererlacher Thomas, Salzer Friedhold, Schieder Pamela, Schwarz Christoph, Trobos Matthias, Turolla Alessandro, Wiesflecker Jakob
指导教师：Verena Rauch, Walter Prenner / 结构顾问：DI Alfred Brunnsteiner / 木作结构：Winkler-Kreutner / 地基处理：STRABAG / 总面积：205m² / 施工时间：2017.5—8
摄影：©Günter R.Wett (courtesy of the architect) (except as noted)

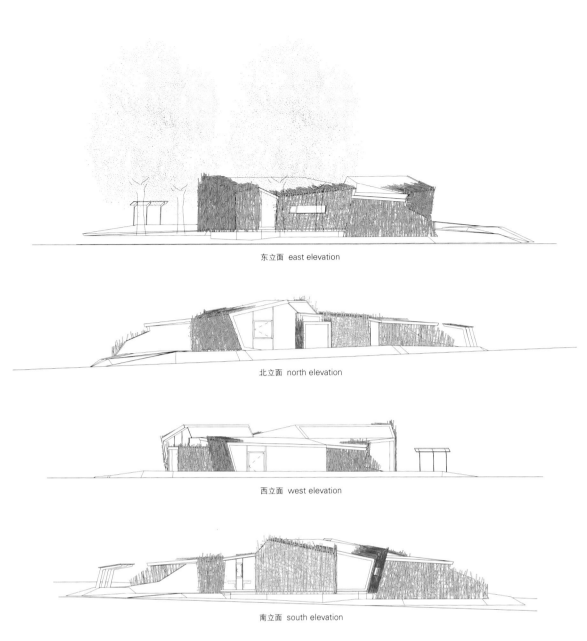

东立面 east elevation

北立面 north elevation

西立面 west elevation

南立面 south elevation

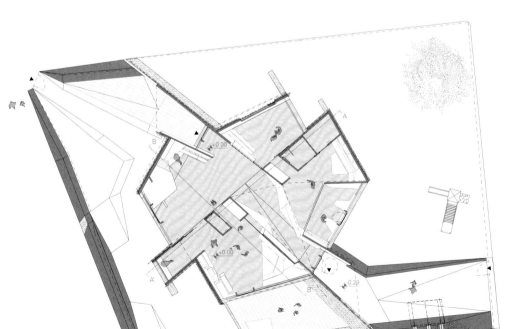

0 2 5m

27

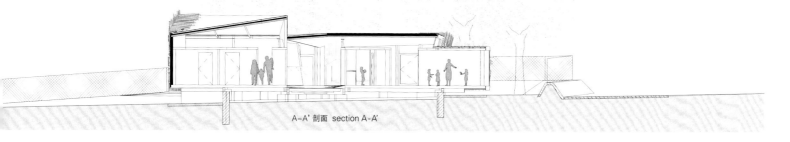
A-A' 剖面 section A-A'

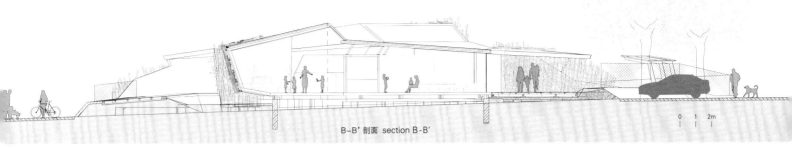

B-B' 剖面 section B-B'

necessary spaces while also enabling the spatial merging of the zones. On the right-hand side, beyond the ramp and past the cloakroom is the office, which is enclosed by glass walls. The adjoining common area for parents is separated from the conference room by the interposed, open infrastructure block. The infrastructure block occupies only about two-thirds of the room's height and includes restrooms, a changing table, a kitchen, and a utility room, all interspersed by passageways. Two classrooms are located on the left-hand side of the building and are flooded with daylight coming in through the floor-to-ceiling windows and skylights. The intersection between the two areas is clearly visible on the ceiling and reflects the spatial continuum. The outdoor area on the river-side is accessible via an IPE terrace that can be reached via the ramp or through the first classroom.

The Spielraeume is purely a wooden structure. In order to minimize production costs, all load-bearing elements were composed of spruce wood cross-laminated timber (BBS). The material also allowed for quick construction of the project by the students, who worked in cooperation with the Winkler company. The nest-like wooden slatted facade envelops the insulated outer walls. The wood in the building interior is untreated. All furnishings were made of three-layer spruce panels and were designed and installed by the students themselves. The interior of the structure is a response to its exterior both in form and in material that composes it.

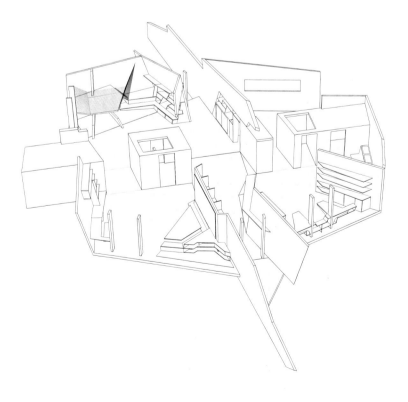

©Matthias Trobos (courtesy of the architect)

©Matthias Trobos (courtesy of the architect)

©Matthias Trobos (courtesy of the architect)

©Matthias Trobos (courtesy of the architect)

- 5mm rubber
- 160mm cross laminated timber
- 240mm glued wood / concrete fundament
- 140mm insulation XPS
- soil

门细部 floor detail

- 21mm terrace decking
- 60x100mm larch-beam
- 100x140mm larch-beam
- neoprene bearing
- 50mm washed concrete slab
- soil

露台细部 terrace detail

- bitumen 2-layered (glued/torched)
- 60mm PU-insulation
- vapour barrier
- 80/120mm cross laminated timber in two drections

屋顶细部 roof detail

- 120mm 3 layers larch-lath
- wind paper
- 60mm PU-insulation
- 100mm cross laminated timber

墙体细部 wall detail

格利法达公立幼儿园
Public Nursery in Glyfada

KLab Architecture

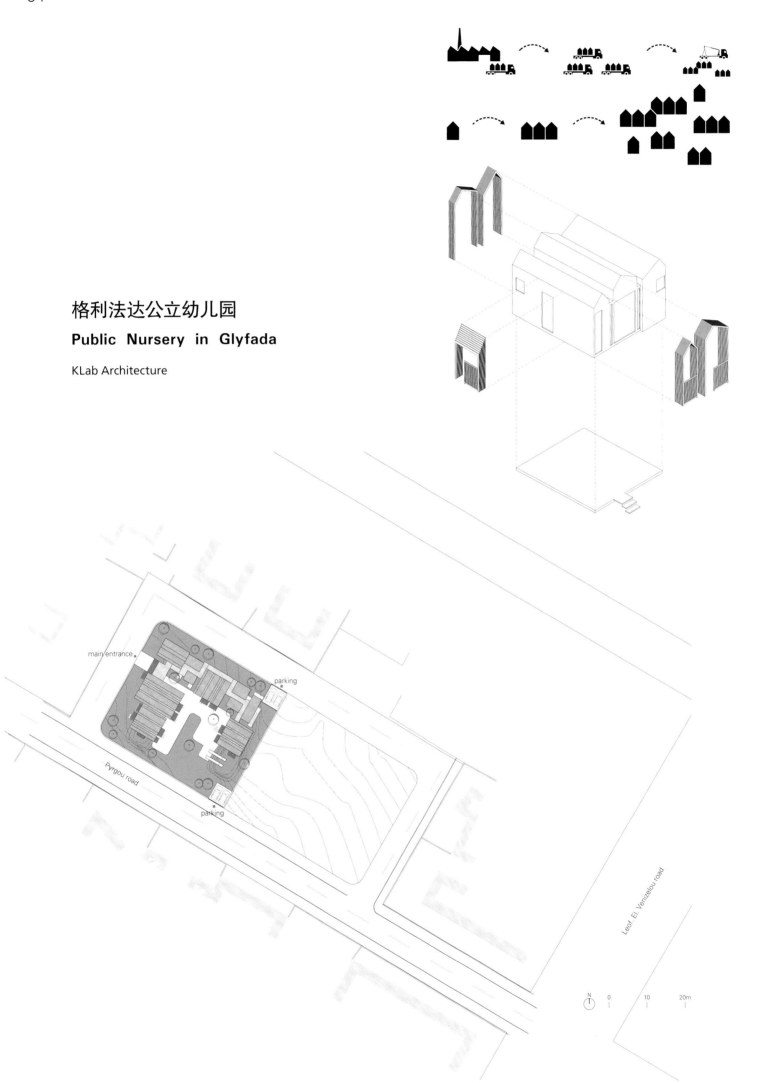

希腊公共建筑的设计和施工一直令人头疼，因为这里的建筑师不能监督施工的情况。本项目所面临的最大挑战之一就是如何重新引入建筑师并使其作为项目的核心贡献者。

项目的另一个挑战则是如何使预制结构成为大众关注的焦点。预制是进行设计竞赛时就提出的条件之一，也将在建筑的最终形态中起到决定性作用。预制的基本单元必须由卡车送往目的地。

项目潜在的基本概念是为使用者进行量身设计，同时还要引进一种新型建筑：都市村庄。建筑主模块的灵感来源于孩子们关于房子的典型画作，教室单元的构成则是经过对基本模块的三重复制得来的。

这所幼儿园的设计保证了所有的教室都有三面可以开放。如此一来，所有教室都被安排在中央庭院的周围，同时又被几个小的中庭隔开。

建设过程中所使用的建筑材料和施工方法虽然都相对常见，但却可以实现更高的复杂度，同时还可以减少能耗。外墙的厚度为10cm，这个厚度有助于实现内部空间的最大化。外墙连同屋顶都已覆盖了保温层。为避免木制绿廊挡住窗户，施工人员对其进行了精心安装，而这正是幼儿园的又一特点，能为孩子们带来持久的舒适感。

设计者还希望绿化能成为此项设计的精髓，尤其是希望内部庭院中繁茂的法国梧桐能为人们带来阴凉。其他树种也将随着季节的变化为项目增添色彩。

The design and construction of public buildings in Greece is problematic, as the architect is not in charge of the construction. One of the biggest challenges of this project was reintroducing the role of the architect as the essential contributor.

Another challenge was bringing prefabrication construction into the spotlight. Prefabrication was one of the conditions of the competition and played a defining role in the final form of the building, as the basic module would have to be transported to its destination by truck.

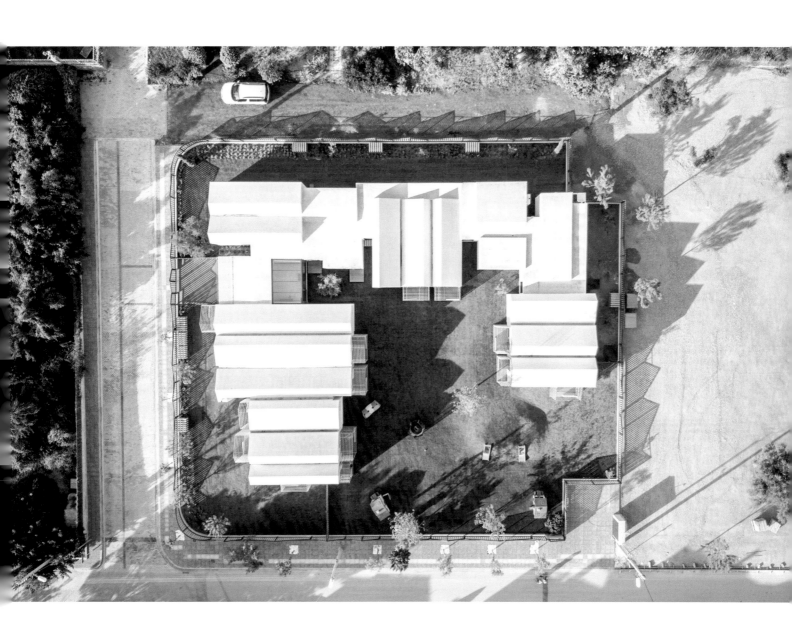

项目名称：4th Public Nursery in Glyfada
地点：Pyrgou str. Glyfada, Athens, Greece
建筑设计：Konstantinos Labrinopoulos – KLab Architecture
设计团队：Konstantinos Labrinopoulos, Veronika Vasileiou, Petra Xynidou-intern, Myrto Lantza-intern
项目管理：Future Constructions
占地面积：1,327.6m²
建筑面积：501m²
总楼面面积：435.7m²
设计时间：2015—2016
施工时间：2016—2017
摄影：©Mariana Bisti (courtesy of the architect)

The underlying concept behind the project was to design a building proportioned and scaled for its users while also introducing a new type of architecture: the urban village. The main module draws inspiration from children's archetypical drawings of houses. The threefold replication of the basic module forms the classroom unit.

The nursery school was designed so that all classrooms have three open sides. As a consequence, the classrooms

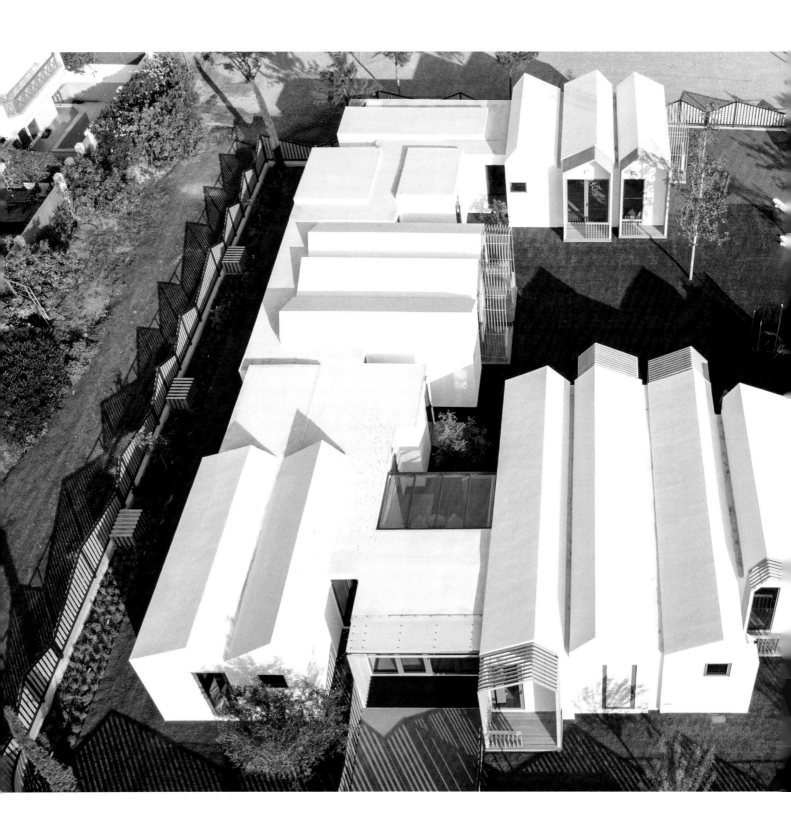

are arranged around a central courtyard while being shaped by the smaller atriums.

The materials and methods used in the construction were relatively commonplace, which allowed for more complexity in the structure as well as a smaller energy footprint. The exterior walls are 10cm in thickness, which allows the maximization of interior space, and they, alongside the roofs, have been clad in exterior wall insulation. The wooden pergolas, which have been carefully placed to avoid blocking the windows, are another feature that will provide children with constant comfort.

The vegetation is expected to become a key element of the design, especially the large platanus tree that will provide shade in the interior courtyard, with other trees that will highlight the changing seasons.

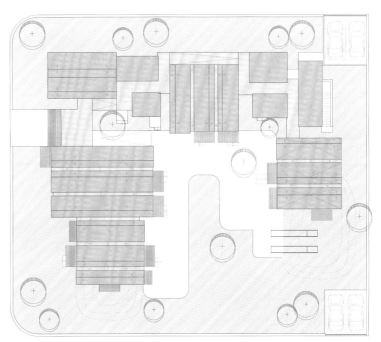

屋顶 roof

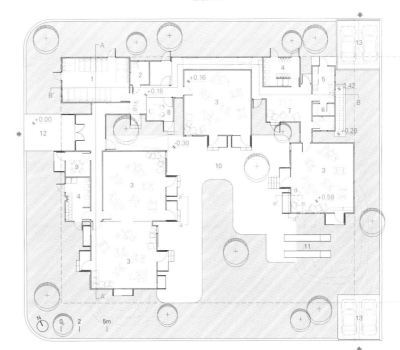

1. 睡房 2. 卫生间 3. 育婴室 4. 卫生间和洗浴区 5. 厨房 6. 库房 7. 多用途房间
8. 行政办公室 9. 教师办公室 10. 游乐区 11. 坡道 12. 主入口 13. 停车区

1. sleeping room 2. toilet 3. nursery room 4. toilet and washing area 5. kitchen 6. storage room 7. multi-purpose room
8. administration 9. teacher's office 10. playing field 11. slope 12. main entrance 13. parking space

一层 ground floor

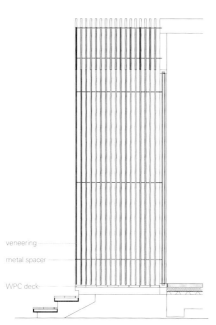

详图b-b' detail b-b'

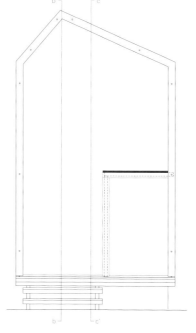

详图a-a' detail a-a'

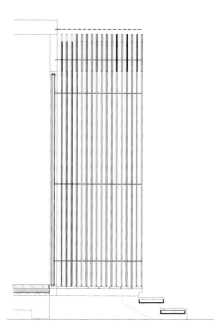

详图c-c' detail c-c'

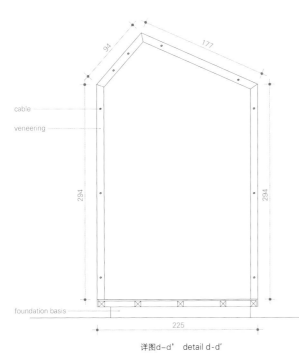

详图d-d' detail d-d'

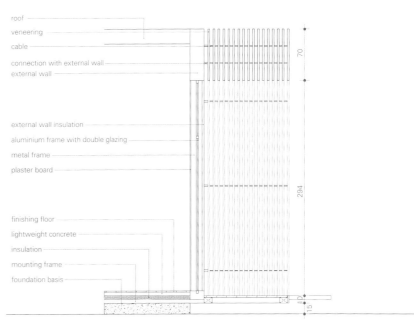

- roof
- veneering
- cable
- connection with external wall
- external wall
- external wall insulation
- aluminium frame with double glazing
- metal frame
- plaster board
- finishing floor
- lightweight concrete
- insulation
- mounting frame
- foundation basis

详图e-e' detail e-e'

西立面 west elevation

南立面 south elevation

东立面 east elevation

北立面 north elevation

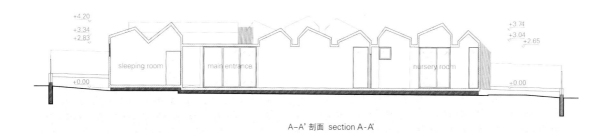

A-A' 剖面 section A-A'

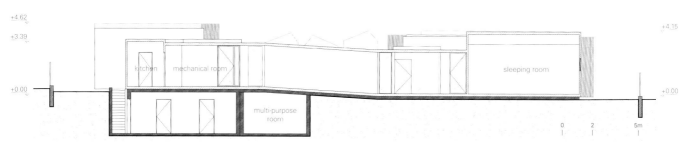

B-B' 剖面 section B-B'

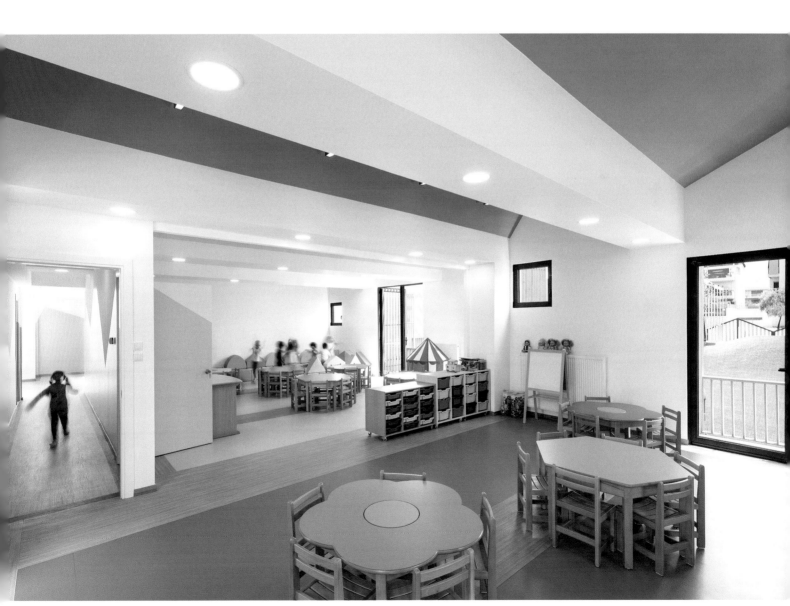

瓜斯塔拉幼儿园
Nursery in Guastalla

Mario Cucinella Architects

马里奥·库奇内拉建筑师事务所 (Mario Cucinella Architects, 以下简称MCA) 在2014年2月举办的建筑竞赛中赢得了冠军, 该竞赛的目的是为瓜斯塔拉区 (位于意大利雷焦伊米利亚) 设计建造一座新的学校。新校舍替代了两所在2012年5月的地震中严重受损的学校, 将招收120名0至3岁的儿童。

以某一教学法为依据, MCA的设计鼓励孩子们与周围环境之间的互动。不论是教育区域的分布状况和建材的选择, 还是室内外的融合设计, 都是基于这种理念出发的。

考虑到与孩子们成长相关的教学和教育两大因素, 新校园进行设计的时候考虑了各种建筑元素, 如建筑内部的形状、组织结构、建筑材料的选择以及所有与光线、颜色、声音、触觉等感官知觉有关的因素。

结构方面, 建筑使用了对环境影响较小的天然材料, 特别是支承结构选用了理想安全的木制框架, 能保证建筑的保温隔热效果。

良好的保温隔热性能、透明墙面的最优分布、先进的雨水收集系统和屋顶嵌入式光伏系统等优势尽可能降低了建筑中机械设备的使用频率, 进而满足了学校的能源需求。这样一来, 学校鼓励学生去探索发现那些复杂却又熟悉的地方, 不同区域特征各异, 以此来提升学生的能力。甚至是教室和实验室之间的连接区域也都有独特的设计, 不仅能满足学生们的好奇心, 还具有趣味性。沿路有宽阔的娱乐和交际区域、可供在之前驻足的壁龛, 以及可以向外看到其他孩子们的活动的透明构件。

除了建筑内部的特色之外, 建筑的外围也确保了感官上的享受。项目利用原有树木将整个校园环境围绕起来, 为学生、老师和家长们的活动营造了一个保护区。

Mario Cucinella Architects won the architectural competition held in February 2014 for the design and construction of the new School in Guastalla District (Reggio Emilia, Italy). The new building replaced two existing schools damaged by the earthquake which struck the territory in May 2012, and will host up to 120 boys and girls between 0 and 3 years old.

MCA aimed to stimulate the child's interaction with the surrounding space according to a vision of "teaching" from the distribution of educational areas and the choice of materials of construction, up to the integration between the indoor and outdoor spaces.

Architectural elements of the new kindergarten – the shape of the interior, their organization, the choice of materials, all

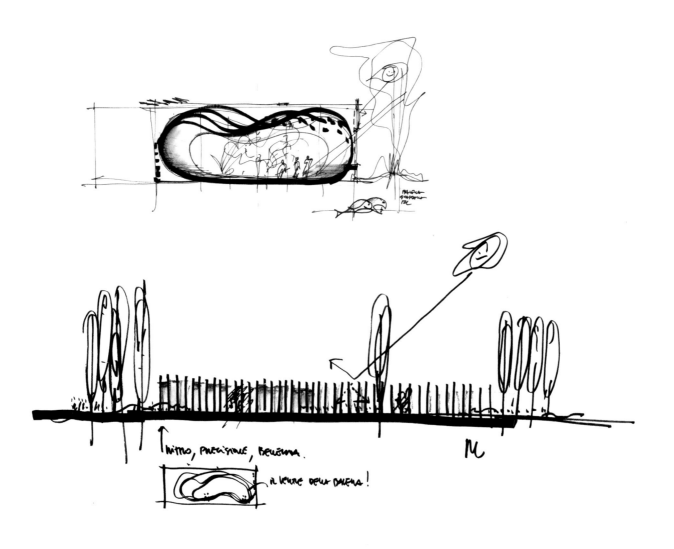

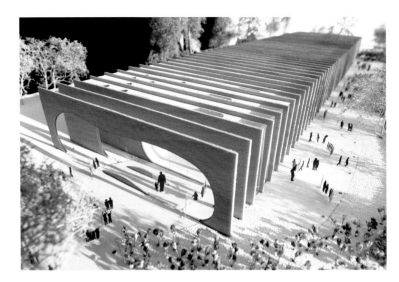

项目名称：Nursery in Guastalla
地点：Guastalla, Reggio Emilia, Italy
建筑设计：Mario Cucinella Architects
项目团队：Mario Cucinella, Marco dell'Agli (architect in charge); Alberto Casarotto, Irene Sapenza, Ferruccio Palumbo, Alberto Bruno, Yuri Costantini (model maker)
景观顾问：Marilena Baggio
客户：Comune di Guastalla
结构：Geoequipe Studio Tecnico Associato
机电设计：Area Engineering srl
声学设计：Ing. Enrico Manzi
建设工程：Scisciani e Frascarelli Impresa Edile
木结构：Rubner Holzbau SpA
供应商：Saitec Company srl
门窗：Promo SpA
建筑面积：1,400m²
单位造价：1,650.00 €/m²
施工时间：2014—2015
摄影：©Moreno Maggi (courtesy of the architect)

the sensory perceptions related to the light, the colours, the sounds, the tactile suggestions – are designed taking into account of the pedagogical and educational related to the growth of the child.

The structure involves the use of natural materials with low environmental impact. In particular, the supporting structure is made of wooden frame: a safe and ideal material to ensure the thermal insulation of the building.

The good insulation, the optimal distribution of transparent surfaces, the use of advanced systems for rainwater harvesting and the insertion of a photovoltaic system on the roof will allow the building to minimize the use of mechanical equipment to meet the energy needs of the school.

Children are driven to discover places that are complex and

at the same time familiar, where they can develop abilities through special features of each. Even areas of connection between the classrooms and the laboratories are designed to be filled with curiosity and pleasure: along the route there are wide play and relationship areas, niches where you can stop and transparent elements to watch out at the activities of other children.

Apart from the internal characteristics, it also articulates the sensory journey outside of the building, which integrates the existing trees and encompasses the structure, creating protected areas for the activities of the children, educators and parents.

NUBO
NUBO

PAL Design Group

NUBO意为"云",正如其意所指,不受拘束、潜力无限。NUBO是一个具有激励性和包容性的儿童娱乐中心,旨在鼓励儿童学习探索,唤醒孩子们无尽的想象力。NUBO尊重孩子们的表现,并不断激励他们,同时也鼓励家长完全参与到与孩子们的互动之中。

NUBO的核心区可充分满足儿童的好奇心,强调在"纯粹娱乐"中去制作、去创造的理念,其中的设施设计巧妙,可灵活地适用于2至8岁的儿童。项目整体设计奉行极简主义,在保障孩子们可以拥有足够的设施能自己发明游戏的同时,移除了不必要的家具和设施,确保处于不同学习阶段的孩子们能安全地探索整个区域。

其次,NUBO还提供"纯粹娱乐"的空间和活动,包括面积广阔、藏书丰富的儿童图书馆,里面有Big Blue Block玩具、Magformer磁力片、乐高Wedo 2.0套件和Kaleido齿轮玩具的多间玩具屋,一间孩子们可以在其中制作各种健康美食的厨房,还有一块玩闹嬉戏专区,孩子们在这里可以滑滑梯、攀爬、捉迷藏。

还有一点很重要,那就是NUBO会鼓励家长和孩子们进行大量的互动,一起度过一段充实美好的家庭时光。NUBO诚邀家长们同孩子们一道来此放松娱乐,并希望家长们也能保持孩童般的好奇心,与孩子们一起学习。毕竟,让每个人都能在这个设计精美的空间内有所享受,才是"纯粹娱乐"的意义所在。

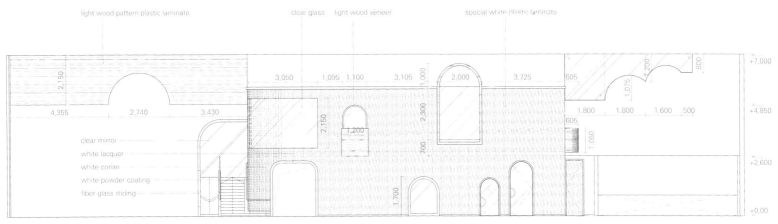

A-A' 立面 elevation A-A'

1. 等候区	10. 咖啡屋	1. waiting area
2. 换鞋区	11. 厨房	2. shoes changing area
3. 接待区	12. 婴童玩乐区	3. reception
4. 图书馆	13. 儿童玩耍塔	4. library
5. 阅读区	14. 卫生间	5. reading stage
6. 艺术室	15. 办公室	6. art room
7. Big Blue Block玩具室	16. 喂养室	7. Big Blue Block room
8. 玩耍台	17. 库房	8. play tower
9. 烹饪站	18. 聚会厅	9. cooking station

10. cafe	
11. kitchen	
12. baby play area	
13. kid's play tower	
14. washroom	
15. office	
16. feeding room	
17. store room	
18. party room	

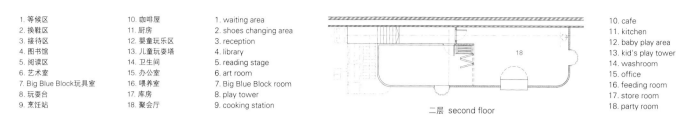

二层 second floor

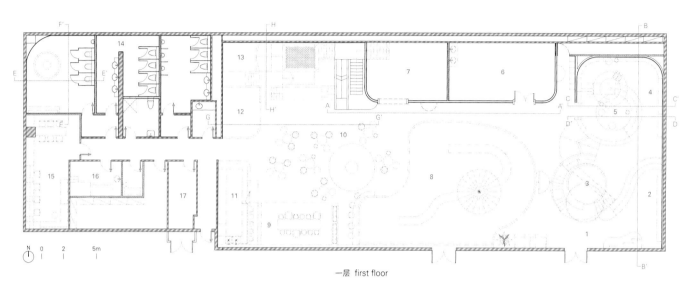

一层 first floor

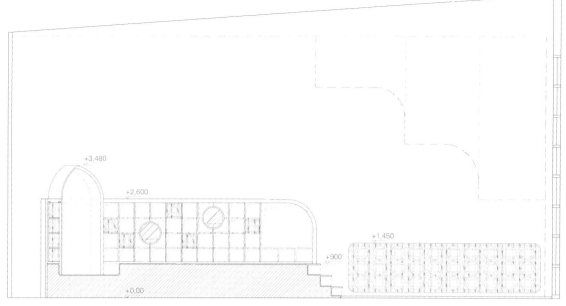

B-B' 立面 elevation B-B'

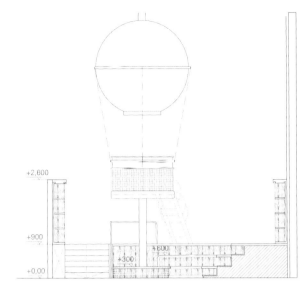

C-C' 立面 elevation C-C'

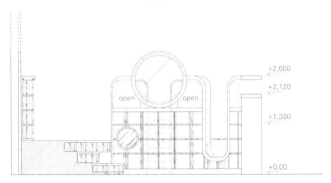

D-D' 立面 elevation D-D'

NUBO – as its meaning 'cloud' suggests – is hard to pin down with its unlimited potential as a stimulating and inclusive play centre to encourage learning, exploration, and boundless imagination. Children are respected yet always stimulated, while parents are encouraged to engage and interact with them through and through.

Its core spaces satiate the kids' curiosity and emphasise the concept of "Pure Play" to make and to create, with its facilities carefully designed and flexibly suited for children aged two to eight. Suited for children in various stages of learning to safely explore the entire space, the overall design takes a minimalist approach to remove unnecessary furniture and equipment – with just enough to invent their own games. In turn, NUBO boasts a variety of 'Pure Play' spaces and activities including an extensive children's library; rooms full of Big Blue Block, MagFormers, Lego Wedo 2.0 and Kaleido Gears; a cafe where children can make a range of healthy dishes; and a zone for active play with opportunities to slide, climb, and hide.

Equally important is to invite parents to spend quality family time together – and with plenty of interactions, too. Adults are also invited to relax and even learn alongside their kids with child-like curiosity. "Pure Play", after all, means something for everyone in this well-designed space to enjoy.

项目名称：NUBO / 地点：Sydney, Australia / 建筑设计：PAL Design Group / 设计师：Joey Ho / 设计团队：Joslyn Lam / 客户：Little Seashell Pty Ltd / 品牌和标志设计：Frost*collective Pty Ltd / 总楼面面积：768m² / 终饰材料：wood, wood veneer, rubber, mosaic, vinyl flooring / 设计时间：2016.1—2016.6 / 施工时间：2016.6—2017.3
摄影：©Amy Piddington (courtesy of the architect) - p.56~57, p.58, p.59 p.60, p.61 bottom; ©Michelle Young (courtesy of the architect) - p.54~55, p.62, p.63; Courtesy of the architect - p.61 top

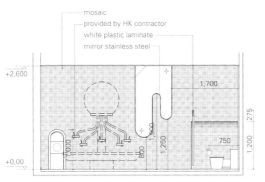

E-E' 立面 elevation E-E'

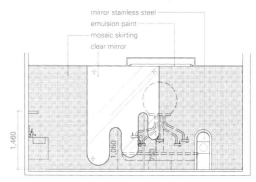

F-F' 立面 elevation F-F'

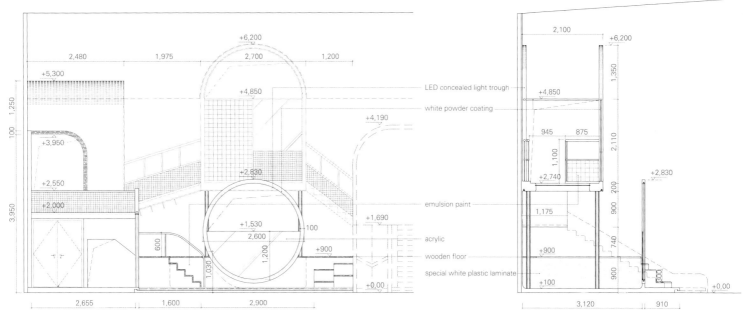

G-G' 立面 elevation G-G'

H-H' 立面 elevation H-H'

Niederolang 幼儿园
Kindergarten Niederolang

Feld72

位于提洛尔南部瓦尔多拉迪索多的Niederolang幼儿园体现了此处缓慢发展的小乡村中传统与现代生活以及自然之间的融合交织。这是Feld72设计的第三所幼儿园，设计实现了它与周围环境的融合，而没有使其受到现有环境的制约。通过巧妙的场地设计，建筑师实现了幼儿园与村子中心的教区教堂及其墓地礼拜堂以及小学等主要建筑的呼应。幼儿园木建筑的四周被一堵实体墙包围，使其仿佛位于一个箱子之内，这种做法为幼儿园提供了一个相对隐蔽的环境。

各种地界和栅栏划定了这所小乡村的结构肌理，这也是一种明确界定私人空间与公共用地的方法。新插入的幼儿园建筑将丰富的边界变化手法体现在其呈围合状态且形式多变的墙体之内，这使其与周围建成环境的主要风格之间形成了一种有趣的文脉融合。建筑外围的墙体在整个村庄结构中形成了明确的空间边界。

嵌入传统文脉中的房屋和庭院组合设计从建筑层面为幼儿园的安全和自由等提供了保证条件。建筑位于整个地块的北端，紧凑的房屋设计为阳光花园留下了一席之地，同时建筑也从功能上将孩子们的这处自由空间与道路分隔开来。从花园的栅栏到园区的围墙，墙体的材质不尽相同，体量也不断变化，既诠释又烘托出现有建筑的设计特点。在庭院内部，墙体则全部采用木制结构，并以一种友好的姿态拥抱着整个花园。

通往建筑的通道上方设有顶棚，以使其免受天气状况的影响。作为一种建筑元素，通道的墙体处理极具趣味性，同时融合了建筑功能和娱乐功能。将墙体这一传统元素与空间结合在一起的做法提升了墙体设计的复杂性和质量感——它邀请孩子们到其中玩耍并为他们提供庇护，不仅能让他们由此看向建筑室内，更可由此看向室外。这一做法还保留了建筑本身干净而又丰富的设计特色。

建材的一致性体现出简洁的特点。幼儿园内部普遍采用的抹灰砖墙和木制结构，为孩子们营造了一种认同感和归属感。得益于对当地木材巧妙复杂的处理，建筑内部营造出一种温馨轻松的氛围。对孩

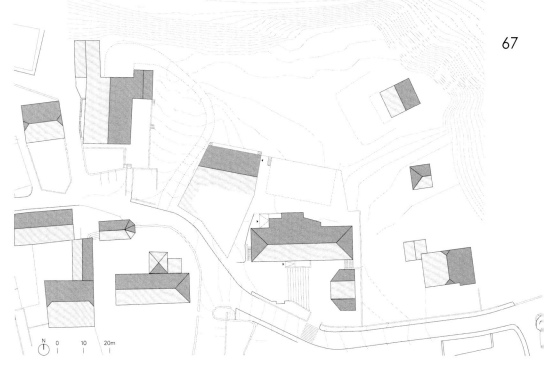
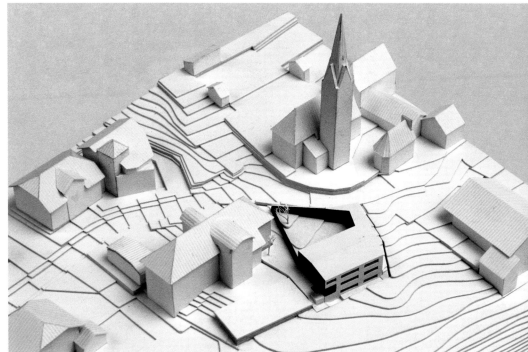

子们来说，这里的各种房间就是他们日常探索的起点。这些房间的设计风格朴实淡雅。宽大的窗台凹进部位和通向衣帽间的步入式隔断家具都可以提供小型的休息区域。会议室和多功能室既可以合在一起使用，也可单独使用。双层空间蕴含的设计理念使得学校的教育方法可以进行各种富有吸引力的变换。这个孩子们和邻里乡亲的相聚之地也正是传统与现代的交融之地。熟悉与独特的风格在瓦尔多拉迪索多的这所建筑内找到了共同的家园。

The kindergarten in Valdaora di Sotto, South Tyrol, illustrates the local mesh of tradition, contemporary life and nature within the slowly grown village structure. This third kindergarten designed by Feld72 aligns itself to the setting without being subordinate to it. Subtly placed, the structure responds to the predominant presence of the parish church with its cemetery chapel and the elementary school in the village centre. The timber building is surrounded by a solid wall, as if it were sitting in a case. Together they provide a hideaway.

Boundaries and fences determine the village fabric. It is a settlement of clear-cut edges separating private and public spaces. The kindergarten as an architectural intervention plays with the boundaries' diverse intensities by putting them into its enclosing multiform wall. That results in an exciting contextual merge with the prevailing element of the built environment. The surrounding wall creates distinct spatial edges within the village ensemble.

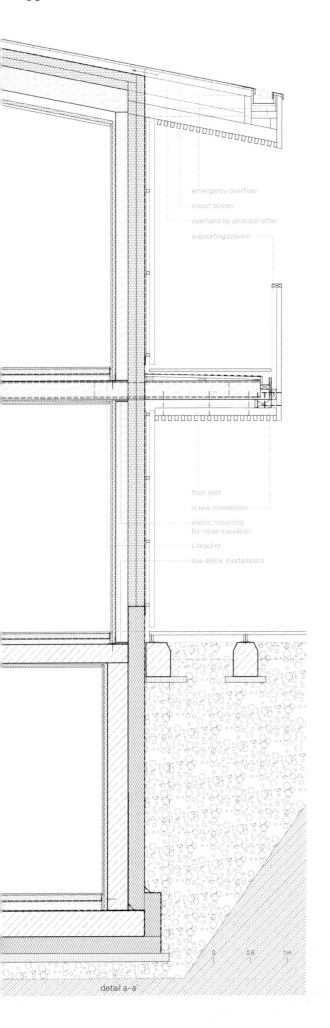

emergency overflow
insect screen
overhand by principal rafter
supporting column

floor joist
screw connection
elastic mounting
for noise insulation
L-bracket
low-shrink mortarboard

detail a-a'

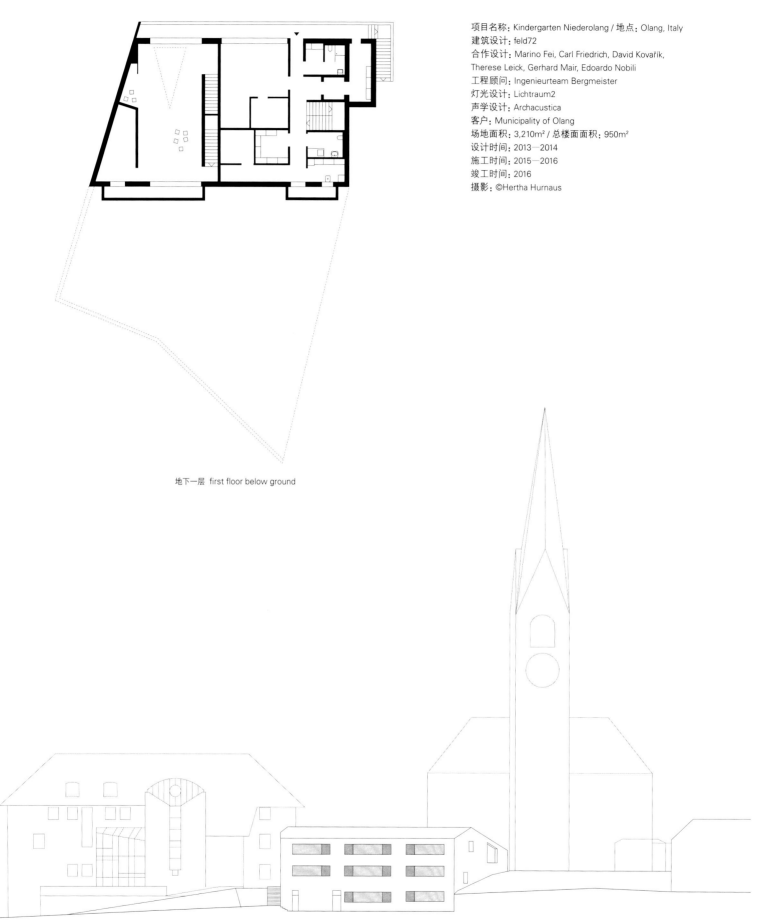

地下一层 first floor below ground

项目名称：Kindergarten Niederolang / 地点：Olang, Italy
建筑设计：feld72
合作设计：Marino Fei, Carl Friedrich, David Kovařík, Therese Leick, Gerhard Mair, Edoardo Nobili
工程顾问：Ingenieurteam Bergmeister
灯光设计：Lichtraum2
声学设计：Archacustica
客户：Municipality of Olang
场地面积：3,210m² / 总楼面面积：950m²
设计时间：2013—2014
施工时间：2015—2016
竣工时间：2016
摄影：©Hertha Hurnaus

北立面 north elevation

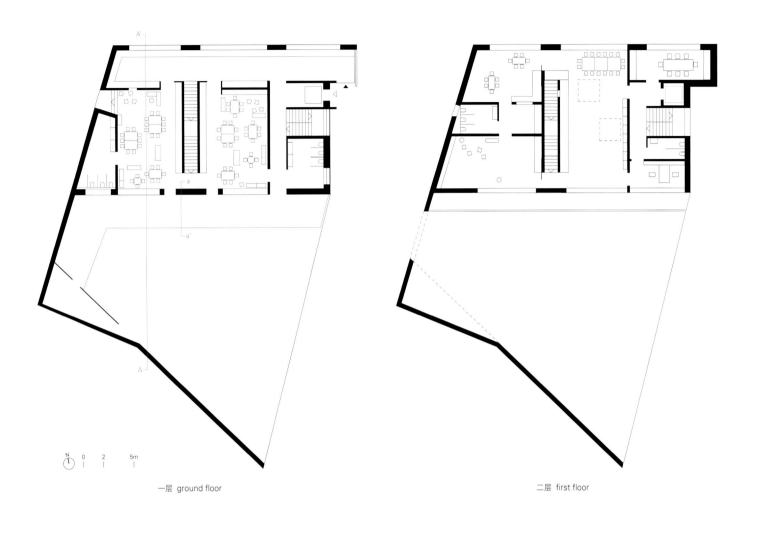

一层 ground floor　　　　　　　　二层 first floor

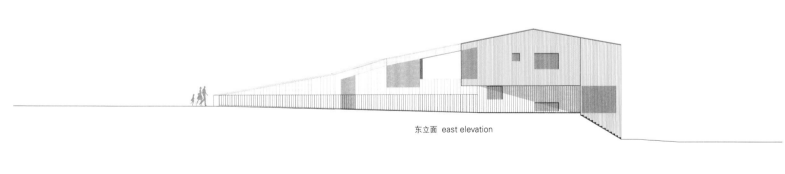

东立面 east elevation

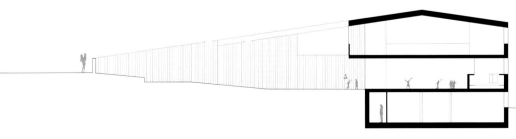

A-A' 剖面 section A-A'

The unity of house and court is embedded in the traditional context and grants the kindergarten the architectural conditions for security and freedom alike. Located on the northern boundary of the property, the compact house offers space for a sunny garden. The free space for the children is functionally separated from the road by the ambient structure. The wall is of varied materiality and volume, since it changes from garden fence to boundary wall, an interpretation as much as an elaboration of the existing. Inside the courtyard, it shows itself entirely in wood and in a friendly gesture embracing the garden.

Roofed, weather-protected areas are set up in the passage to the building. The handling of the wall as an architectural element unfolds in a playful way, architecture and playground fused. The traditional element of fencing-in a property gains in complexity and quality – it shelters and invites children to play here and provides insights and outlooks. The building itself remains clear and yet manifold.

Uniform materiality displays simplicity. Plastered brickwork and wood prevalent inside the kindergarten create identity and a sense of belonging. A warm and easy atmosphere is generated within the interiors thanks to the subtly complex use of local timber. For the children, the group rooms represent the starting point of their daily explorations. The rooms are designed to be unpretentious and calm. Small areas for retreat are offered by the large window recess or the accessible partition furniture leading to the cloakroom.

The assembly room and multi-function rooms can be used connected or individually. On both levels, the suggested spatial concept enables attractive variations in the educational approach. Here, a place for children and the village community, is the interface for tradition and modernity. The familiar and the exceptional together find a home in Valdaora di Sotto.

ATM 幼儿园
ATM Nursery

Hibinosekkei + Youji no Shiro

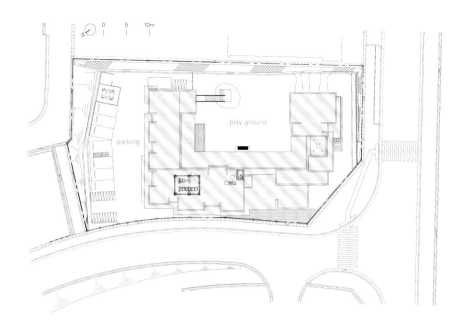

ATM幼儿园位于日本千岁新城，横跨丰中市和吹田市，该地段的前身为消防局。这一地区的社区最初是围绕住宅区发展起来的，但最近开始减少，有的甚至完全消失。

幼儿园是孩子们性格发展的地方，了解自己的成长环境和其所在社区的历史至关重要。在开展此项目的过程中，Hibinosekkei + Youji no Shiro对这里的旧式住宅小区进行了重新诠释，并将幼儿园规划成当地社区的中心。

这所幼儿园的外部大胆摒弃了现有住宅小区千篇一律的瘦长型设计手法，其错层设计以一种拼接的方式为老旧的社区注入了一股生机。建筑的外围设有阳台，而阳台也同时是这里老式住宅区的标志性特征之一。阳台上装配了球网、攀爬设备、坡面、长椅和供儿童攀爬的架子，让孩子们在玩耍的同时也能锻炼身体。宽阔的阳台可将孩子们的活动展现给社区邻里。

在这里，孩子们有时也会遭遇挑战的失败或者受点轻伤，但这也能激励孩子们取得进步，迎接挑战。

项目的主要目标之一是鼓励人际沟通。为达到这一目的，建筑的设计可使使用者们在任意方向上都拥有清晰的视野。例如，幼儿园的厨房和餐厅都面向街道和室外，同时又与外部走廊和玩耍露台相连。这样一来，不仅当地的居民可以时刻关注着孩子，老师们也能更容易地观察到外面的情况。

幼儿园内的每一间育婴室都配有大玻璃窗，视野开阔，能看见玩耍露台和走廊，以此鼓励不同年龄段的孩子们在一起玩耍。这样一来，小孩子们就更加愿意和大孩子们在一起玩耍，并渴望成长；大孩子们则会自然而然地对小孩子产生责任感，并学会友善待人。

出于对安全因素的考虑，许多幼儿园开始变得越来越封闭，但这家幼儿园却通过对外开放和培养当地居民及园内人员的社区意识来提高安全性。

露台是当地居民与园内人员交流之地。这种积极的互动可以帮助孩子们学会沟通，感受社区的温暖。多年后，孩子们长大成人，他们自己也会成为这个曾养育他们的社区中的一员，并使社区变得更加美好。

Built on the site of a former fire department, this nursery is situated in Chitose new town, which straddles the border between Toyonaka city and Suita city. Communities in this area originally sprang up around residential complexes, but have recently begun falling into decline and disappearing altogether.

A nursery is a place where a child's character develops, and it is crucial for children to understand their historical surroundings and the history of their own neighbourhood. In developing this project, Hibinosekkei +Youji no Shiro reinterpreted the old residential complexes and planned the nursery as a hub for the local community.

The nursery exterior is a bold departure from the uniformly long, thin design of established residential complexes, its mismatched levels bringing a patchwork splash to the old neighbourhood. The balcony at the periphery of the building, one of the defining characteristics of the old residential complexes, has been equipped with nets, climbing equipment, a slope, a bench, and a monkey bar so children can both play and exercise at the same time. The openness of the balcony puts the children's activity on display to the neighbourhood. This area also allows children to fail sometimes or to be lightly injured, helping them to build a sense of advancement and challenge.

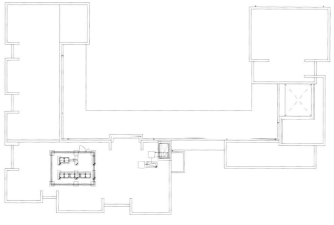

屋顶 roof

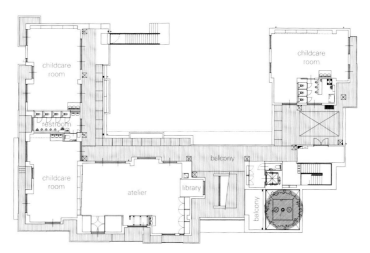

二层 first floor

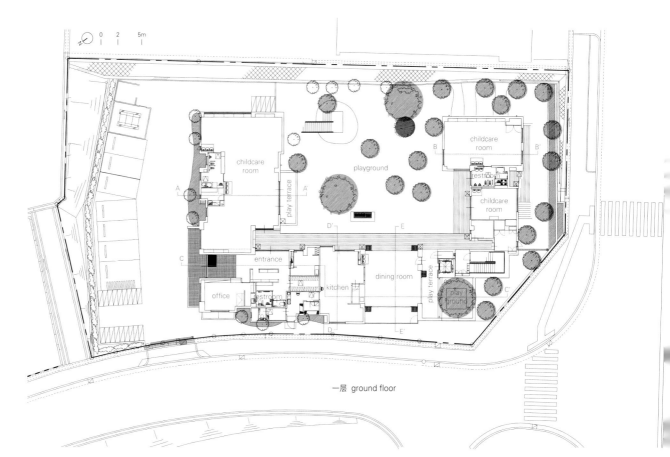

一层 ground floor

西立面 west elevation

南立面 south elevation

操场处的南立面
south elevation at playground

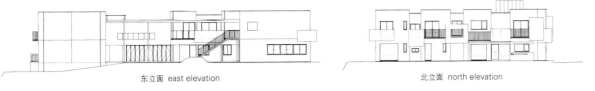

东立面 east elevation 北立面 north elevation

操场处的北立面
north elevation at playground

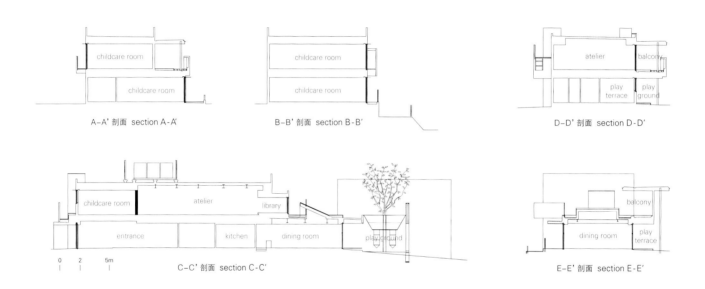

A-A' 剖面 section A-A'   B-B' 剖面 section B-B'   D-D' 剖面 section D-D'

C-C' 剖面 section C-C'   E-E' 剖面 section E-E'

项目名称：ATM Nursery / 地点：Toyonaka, Osaka, Japan / 建筑设计：HIBINOSEKKEI, Youji no Shiro
场地面积：2019.7m² / 建筑面积：694.87m² / 总楼面面积：1080.3m² / 建筑规模：two stories above ground
结构：steel frame / 竣工时间：2017. 3 / 摄影：©Studio BAUHAUS (courtesy of the architect)

Encouraging human contact was one of the main goals of this project. To that end, the structure was designed so that occupants would have clear lines of sight in every direction. For instance, the kitchen and the dining room are open to the street and the exterior, meanwhile they are also connected to the outside corridor and play terrace so that local residents can keep an eye on the children and teachers will be able to more easily see what is happening outside.

Every nursery room has large windows that open to the side of the play terrace and the corridor so that children of different ages will be prompted to engage with one another. As a result, younger children grow attached to older children and become more eager to grow up. Older children, for their part, develop kindness through their responsibility for the younger children.

Many nurseries have become more and more closed for security purposes. However, this nursery enhances security by opening its doors to the outside world and fostering a sense of community among local residents and within the nursery itself.

The terrace is where local residents can communicate with those in the nursery. The buildup of positive interaction will help the children learn to communicate as they engage with the warmth of the community. In the future, these children will grow and themselves become the community that raised them, making their society a better place to live in.

Chuon Chuon Kim 第二幼儿园
Chuon Chuon Kim 2 Kindergarten

KIENTRUC O

越南建筑工作室KIENTRUC O在胡志明市设计建造了一所幼儿园,以各种堆叠的砖砌体量为特色,各体量之间以悬臂结构和坡道相连。KIENTRUC O认为应为全体师生建造具有激励精神氛围的教育类建筑,鼓励孩子们进行探索,并为他们提供丰富的互动机会。幼儿园想要向人们徐徐灌输一种开放式思维方式,并夹杂些许好奇的火花,鼓励不同年龄的人们都能在轻松宁静的氛围中进行冒险和探索。

整个幼儿园就像一座大型的乐高建筑,完全由裸露的砖块垒砌而成,具有充满趣味的图案和开口,不仅展现出独特的审美观,还有利于建筑的自然通风。建筑师们在明暗的协调方面采用了精心的设计。不同空间的风格有着很大的反差,突出了整个幼儿园建筑之旅的奇妙。

教室等房间围绕一个娱乐核心区分布。楼层平面交错排布,以促进上下层间的垂直互动,使孩子们更容易接受周围的环境,并激发其内在创造力。

和教室的安静氛围相比,娱乐核心区充满活力。以一层花园为起点,连续空间形成的孔洞可使人们从外部看到内部,或者是从内部看到外部。再往上,室内空间最终与屋顶的开放式花园连接在一起,站在这里可以俯瞰整条Saigon河,让人流连忘返的美景就在那里等着你去发现。

在幼儿园里可以进行一次非常自由的探索之旅,随着场地的不断变化,你能感受到每个定制空间所带来的无尽感受与体验。总的来说,幼儿园处处充满惊喜,不论是对孩子还是对成人来说,这里都会让他们难耐好奇。

The Vietnamese architecture studio KIENTRUC O designed a kindergarten in Ho Chi Minh City characterized by piled-up brick volumes, with cantilevers and ramps connecting them. KIENTRUC O believes in creating educational buildings that provide stimulating environments for both pupils and staff with architecture that promotes exploration and offers various opportunities for interaction. Instilled within the kindergarten is an openness with a spark of curiosity that allows people of all ages to venture and explore the space in a relaxing and calming atmosphere.

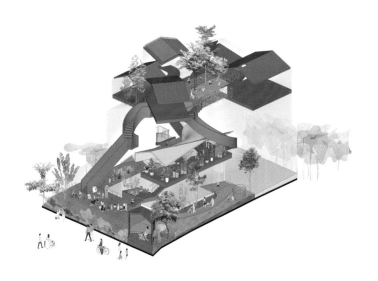

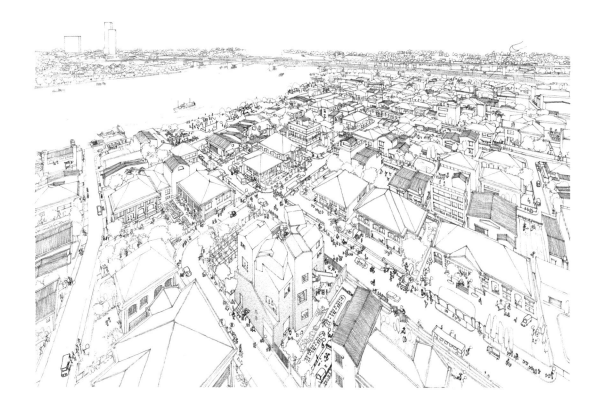

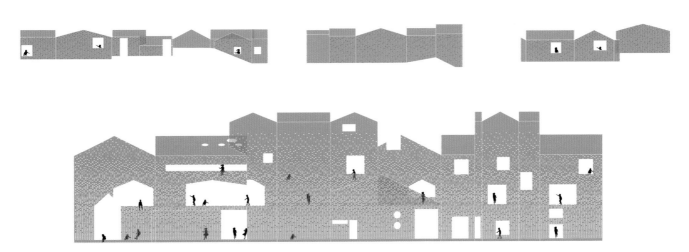

砖墙立面展开图 brick facade planar figure

项目名称：Chuon Chuon Kim 2 Kindergarten / 地点：District 2, Ho Chi Minh City, Vietnam / 建筑设计：KIENTRUC O
主创建筑师：Đàm Vũ / 项目团队：Anni Lê, Tài Nguyễn, Dân Hồ, Phương Đoàn, Duy Tăng, Giang Lê, Đức Lê, Tân Phạm / 承建商：Đinh Đức Anh Vũ
客户：Chuon Chuon Kim Edu / 总楼面面积：409m² / 竣工时间：2017 / 摄影：©Hiroyuki Oki (courtesy of the architect)

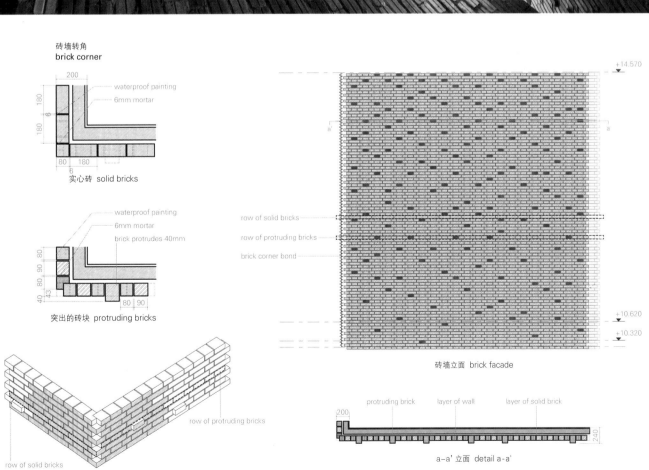

砖墙转角 brick corner

实心砖 solid bricks

突出的砖块 protruding bricks

row of protruding bricks
row of solid bricks

row of solid bricks
row of protruding bricks
brick corner bond

砖墙立面 brick facade

protruding brick | layer of wall | layer of solid brick

a-a' 立面 detail a-a'

A-A' 剖面 section A-A'

B-B' 剖面 section B-B'

Like a giant Lego building, the kindergarten is constructed entirely in bare brick-forming patterns and openings that look playful, conveying a unique aesthetic value and promoting natural ventilation. The architects take a thoughtful approach toward negotiating light and dark. The spaces they create have tremendous contrasts, accentuating wonderful journey through the school.

Classrooms and some other rooms are organized around a playful core. Each floor is arranged in an alternating pattern to enhance vertical interaction, encourage children to be more receptive of their surroundings, and stimulate their inner creativity.

Juxtaposed to the calming atmosphere of the classrooms, the core is full of movement. From the garden on the ground floor, the spaces form an aperture that frames a continuous perspective that is visible from outside in and inside out. Continually upward, the interior spaces connect to an open rooftop garden, awaiting to be discovered with a rewarding experience of the infinite vista of the Saigon river.

The journey of discovery in the kindergarten is a very liberating one because of the continuous changes, and the endless experiences are tailored particularly to each space. Conclusively, Chuon Chuon Kim 2 Kindergarten is a place of surprises that will never cease to tickle the curious souls, children and adults alike.

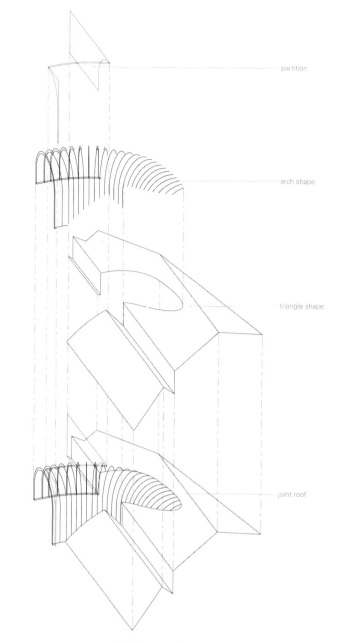

中央上空区 central void

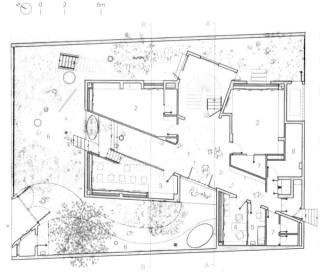

1. 大堂 2. 教室 3. 办公室 4. 主任室 5. 儿童护理室 6. 操场 7. 洗手间 8. 库房
1. lobby 2. classroom 3. office 4. director room
5. childcare room 6. playground 7. restroom 8. storage
一层 ground floor

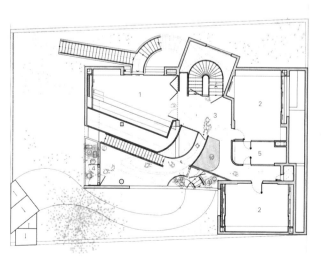

1. 艺术室 2. 教室 3. 大厅 4. 花园 5. 洗手间
1. art room 2. classroom 3. lobby 4. garden 5. restroom
二层 first floor

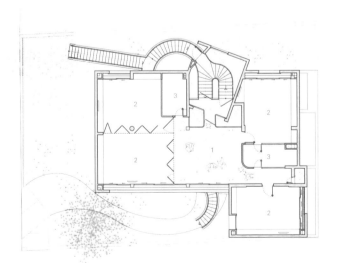

1. 大厅 2. 教室 3. 洗手间
1. lobby 2. classroom 3. restroom
三层 second floor

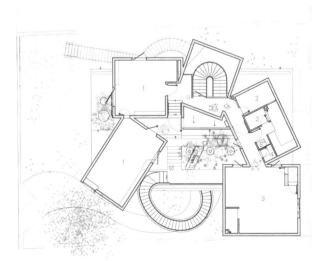

1. 艺术室 2. 洗手间 3. 厨房 4. 库房 5. 洗衣房
1. art room 2. restroom 3. kitchen 4. storage 5. laundry room
四层 third floor

冒险城堡

KIENTRUC O访谈录

Adventure Castle

Interview with KIENTRUC O

C3：参考了Chuon Chuon Kim第一幼儿园的设计后，Chuon Chuon Kim第二幼儿园继续采用这种房屋式积木的建筑风格。但不同的是，第一幼儿园是将带有坡屋顶的体量放置于常规的建筑表皮之中，而第二幼儿园的设计则更为大胆，直接将各个体量多变的角度和缩进设计都暴露在外。

KIENTRUC O：孩子们的身高决定了他们感知空间的视角。因此，当空间的尺寸与他们的身高相符时，就能创造出既安全又具吸引力的环境。这两所幼"玩"园的塑造和打造都基于这一基本理念，并将其用坡屋顶建筑的形式表达了出来。一个简单的三角几何图形因其不同的高度就可呈现出动态的空间布局效果。

尽管两所幼儿园的设计理念一致，但为了将二者区分开来，我们还是采用了两种完全不同的建筑组合方法。第一幼儿园采用的是空间嵌套结构，而第二幼儿园则基本是将被组合成一个独特的房屋式几何体的多个房屋式实体砌块用几何形式表达了出来。对立体几何的运用则是为了锁定整个大型建筑中被围合之后形成的无定形孔洞空间，这种空间是一种对孩子们的生活之地和幼儿园的所在之处的直接环境的建筑表达。

在立方体的两侧打通了多个出入口，有意识地将内部建筑与外部自然环境连通，凸显儿童与大自然关系的重要性。另外，这个孔洞空间中也因分布和点缀有大小、形状和高度各异的更小型孔洞空间而显得富有趣味性，从而创造出丰富的空间体验。整个建筑更像是一个洞穴式建筑，只不过通过运用抽象几何的方式进行表达。幼儿园像一个令人好奇的盒子，充满了空间趣味性，能激励孩子们去冒险、去探索。

C3：从幼儿园倾斜设计的入口开始，窗户、开口和楼梯也都被设置在一个略微扭曲的网格上，似乎一直在鼓励园内的孩子们从多个视角去发现外部的世界。您这种有趣的网格设计灵感是从何而来的呢？

KIENTRUC O：这两个幼儿园都是翻新项目，之前的建筑布局也都十分生硬。起初，我们根据之前的网格系统，按一般顺序对两所幼儿园的空间进行了组织。随后，我们分析了周围的环境，并引进了一种全新的、带角度的空间转换方案。新方案所形成的动感激发了孩子们的内在好奇心，这个结果出乎预料。这些开口，再加上与之形成互补关系的竖向楼梯，无论是从内部还是从外部，都给人带来一种旅途般的体验。我们为每所幼儿园都量身打造了这种独特的空间解决方案。

C3 *With reference to Chuon Chuon Kim Kindergarten 1, Chuon Chuon Kim Kindergarten 2 continues to play with "house-like" blocks. However, instead of "casing" shed-roofed volumes in a regular envelope, the kindergarten 2 decides to expose the bold volumes of varied angles and indents.*

KIENTRUC O Children perceive space from a perspective relevant to their heights. Therefore, when space is suited for their heights, it creates an environment that is safe and inviting to inhabit. Both Chuon Chuon Kim kinder"play"gartens are sculpted and shaped by this fundamental understanding architecturally expressed out through the form of a shed roof. A simple triangular geometry offers dynamic spatial setting due to its different heights.

Although based on the same principle, Chuon Chuon Kim 1 and Chuon Chuon Kim 2 are methodically composed differently to distinguish the two projects. Chuon Chuon Kim 1 is a space within a space, whereas Chuon Chuon Kim 2 is a literal geometrical expression of multiple house-like solid blocks morphed into a single distinctive house-like geometry. The use of solid geometry is to capture the formless void encased within the large block. The void is an architectural reflection of the immediate context where the children are living and where the Chuon Chuon Kim 2 kindergarten is located. It carves out openings from the solid on two sides to establish a purposeful connection with nature, underlining the important relationship between children and nature. In addition, the void is made more interesting by accumulation of smaller voids of various sizes and shapes connected at different heights, hinting diverse spatial experiences. That is a setting which can be assimilated to cave-like architecture, only to be appropriated through the use of abstract geometry. The kindergarten is a curious case of spatial playfulness that stimulates children's sense of adventure and exploration.

C3 *From the entrance placed obliquely, many elements like windows, openings, and spiral staircases are set on a slightly twisted grid. It seems that it constantly encourages the kids inside to frame the outside world from many perspectives. Where did you get inspiration for this playful grid?*

KIENTRUC O Both kindergartens are renovation projects whose existing structural grids were very rigid and stiff. Initially, we organized the spaces for Chuon Chuon Kim 1 and 2 in a general order according to the existing grid system. Then, we analyzed the surrounding context and introduced a new method of angled spatial alteration. Consequently, this created unexpected movements that stimulate children's inner curiosity. The openings, complemented by the vertical movement of the stairs, make a trip-like experience highly noticeable from both inside and outside. A unique spatial solution is tailored to each Chuon Chuon Kim Kindergarten.

C3 *When brick is mainly used for exterior finish, the building*

C3: 当砖块被用于外墙终饰的时候，一般来说建筑都会在整个环境中略显突兀。但不知为何，这个项目却巧妙地回避了这一点。您在这一方面采取的具体策略有哪些？

KIENTRUC O: 即使是单一色调的砖块也能创造出多元有趣的建筑效果。一开始，我们想采用中空立方体砌块的形式来容纳各项规划要素并使建筑呈现出几何感。但是，为了使其不论是从内部还是从外部都看起来柔和、友好，建筑还要做到足够通透。为了达成这样的目标，我们对建筑的通透度进行了精心调整：建筑底部开放度很高，越向上开放度越小，这样做不仅能使它在视觉上更为柔和，还有助于通风。

此外，当对建筑空间进行组合时，裸露在外的砖块与灰浆的接缝可使人们清晰地感知到建筑的结构，这和孩子们搭建他们的乐高积木房子是一样的。

C3: 这个幼儿园中的游乐场地或大或小，形成了一个"空间集群"。它是一个"动若脱兔，静若处子"的建筑，对此您有何看法？

KIENTRUC O: 我认为，如果一栋建筑能同时兼具复杂性和简约性，那么这座建筑就是非常有趣的。说到第二幼儿园的虚实设计，其外部立方体结构为实，设计为简，并以坡屋顶的形式呈现；其内部的各种孔洞设计为虚，并正好形成了空间体验的复杂性。这是一个处于祥和环境中的质朴的积木式建筑，其主入口面向一个安静的花园而开放。

C3: 幼儿园最突出的特点之一就是楼梯与顶部像房子一样的盒子之间的连接。这是让孩子们用来探险的吗？这种感觉就像沿路上山，等我们到山顶的时候，就能以平静的心境享受周围的环境。

KIENTRUC O: 孩子们通常都喜欢攀爬，因为攀爬能使孩子们更深刻地理解他们赖以生存的世界，也能激发他们无穷无尽的想象力。特别是那些非常规用途的空间，更让孩子们兴奋不已。第二幼儿园的空间设计正是基于这种理念。这些空间从一些像房子般的小砌块中雕刻出来，形成了一个能为身在高处的孩子们提供安全区域和平和感的空间，同时也能将这座城市中最大河流的景色呈现在孩子们的眼前。

有趣的是，根据远景法则，当孩子们在屋顶上的时候，他们所看见的附近房子的比例都和幼儿园屋顶相似。因此，孩子们能从他们的视角去了解附近的环境，并构想他们自己眼中的世界。

may look so bulky as to overwhelm the surroundings. But somehow this project skillfully washes away such concern. What was your specific strategy upon this?

KIENTRUC O The monochromatic brick tone creates diverse and interesting architectural outcomes. From the beginning, we wanted to use a solid block that is hollow to house the programmatic elements and give the building a geometrical impression. However, it had to be porous enough to look soft and friendly from both outside and inside. To achieve this, the building's porosity is fine-tuned; it opens out at the bottom and gradually closes up to the top. This effect makes the building look softer at eye levels and facilitates ventilation better.

In addition, exposing the brickwork and revealing the mortar joints give a sense of construction clarity when the building is assembled, just as how a child constructs his Lego house.

C3 *The kindergarten gives a feeling of a playground with large and small gestures, making a cluster of "spaces". What was your consideration on "alive and breathing spaces to flow" and "calm and secret spaces to stay"?*

KIENTRUC O I think a building is interesting when it can accommodate both complexity and simplicity at the same time. Speaking of solid and void at Chuon Chuon Kim 2, the internal voids make up the complexity of the spatial experience within the simplicity of the external solids, expressed in a form of a shed roof house. This is a modest building block in a peaceful setting, whose main entrance opens directly to a quiet garden.

C3 *One of the most prominent elements in the kindergarten is the linkage of the staircases and the house-like boxes on the top. Are they to give the chance of adventure to the children, just like the way up to the mountain and then the top of it which we finally reach and which lets us enjoy the surroundings in peace of mind?*

KIENTRUC O Children often like to climb on objects because they get larger viewpoints of the world they inhabit in, and a stimulation for their endless imagination. Especially, spaces that are not meant for normal uses always tickle their excitement. Spaces in Chuon Chuon Kim 2 Kindergarten are created based on this understanding. These spaces are sculpted by small house-like blocks, forming a safe zone that offers a sense of peace for the children when they are high above the ground but at the same time, open up a view to the city's largest river.

What is interesting when children are up on the roof is that the houses nearby have similar proportions to the kindergarten roof according to the law of perspective. Hence they can comprehend their viewpoints to the nearby surroundings and draw up their own picture of the world.

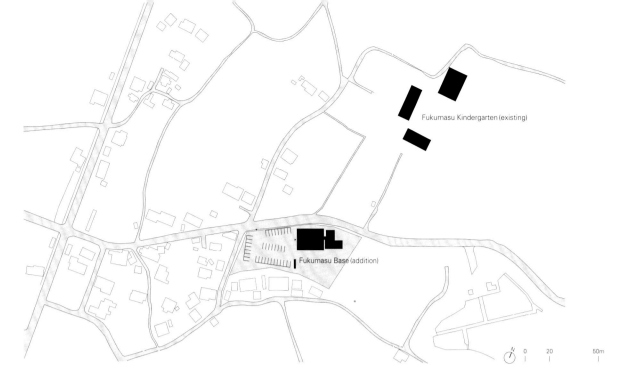

Fukumasu 基地
Fukumasu Base

Yasutaka Yoshimura Architects

Fukumasu基地是日本千叶县一家现有私立幼儿园的扩建项目,是一个为儿童及其父母设计的全新活动场所,此地距离东京有一个半小时的火车车程。该项目的设计师是Yoshimura,其设计体现了"局部独立主义"的理念。他尽可能地将设计过程中的尝试和失误展现给公众,以期在项目竣工之时能顺利地将该建筑的主观意义传达给建筑的使用者。举例来说,建筑师引入了预制薄膜仓库来替代现存旧式仓库,而设计初期曾计划要对此仓库进行重复使用。还有室内,后期出于结构上的需要,建筑师将带有金属节点的斜梁安装到了薄薄的墙壁上,而实际上,只要增加墙体的厚度就完全能保证其稳定性。

幼儿园的负责人认为:不应在一开始就设定好建筑的用途,而应由使用者自行决定。设计过程中,思路的不断明朗和实现正是对以上理念的回应。此外,不隔断任何房间的可折叠式隔墙也是对负责人所秉持理念的一种回应。由于一开始就没有将房间命名,所以以墙体围合而成的各区域的用途能不断地引发学生的思考。通过建筑师和业主间的完美合作,该设施已成为一个由私人拥有并运营的公共建筑。

Fukumasu Base is a new facility supporting children and their parents, which is added to an existing private kindergarten located in Chiba, about one and a half hours by train from Tokyo. Yoshimura designed it with his concept "Ad-hocism" by making the trials and errors during the design process visible as much as possible so that the subjectivity of the architecture moves smoothly from the architects to

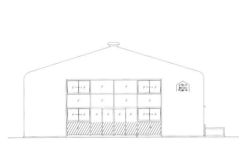

西立面 west elevation

东立面(帐篷封闭)
east elevation (covered tent)

南立面 south elevation

东立面 east elevation

北立面 north elevation

the users of the building when the construction work is completed. For instance, the readymade membrane warehouse introduced by the architect was the replacement of the existing old warehouse which was supposed to be reused at the beginning of the design process. Regarding the interior, the diagonal beams with metal joints were added to the thin walls afterwards when it was needed structurally although it could also be stable by making the walls thicker. The chair of the kindergarten believes that the way of using the building should not be given primally and the users themselves should commit and determine it. Clarification and visualization of the traces of thoughts during the design process could be the opportunity to answer to it. Furthermore, folding walls without closing any rooms off are also a reaction to the chair's belief. The corners created with walls constantly stimulate the children to think how to use them because there are no rooms named from the beginning. Through the ideal collaboration of the owner and the architect, the facility has become a public building owned and run by a private sector.

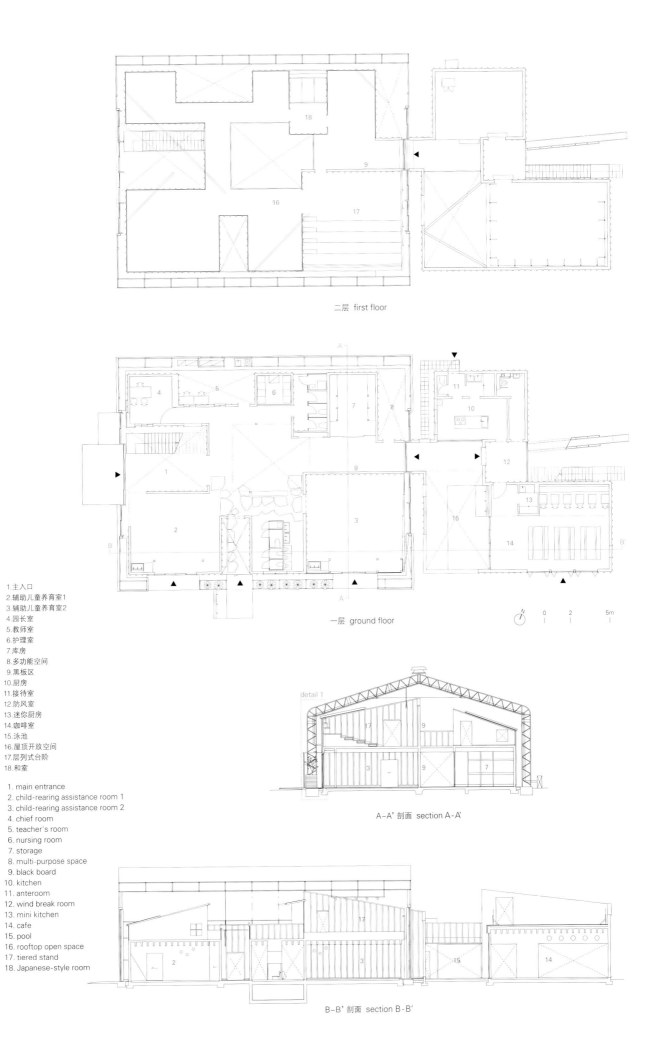

二层 first floor

一层 ground floor

1. 主入口
2. 辅助儿童养育室1
3. 辅助儿童养育室2
4. 园长室
5. 教师室
6. 护理室
7. 库房
8. 多功能空间
9. 黑板区
10. 厨房
11. 接待室
12. 防风室
13. 迷你厨房
14. 咖啡室
15. 泳池
16. 屋顶开放空间
17. 层列式台阶
18. 和室

1. main entrance
2. child-rearing assistance room 1
3. child-rearing assistance room 2
4. chief room
5. teacher's room
6. nursing room
7. storage
8. multi-purpose space
9. black board
10. kitchen
11. anteroom
12. wind break room
13. mini kitchen
14. cafe
15. pool
16. rooftop open space
17. tiered stand
18. Japanese-style room

A-A' 剖面 section A-A'

B-B' 剖面 section B-B'

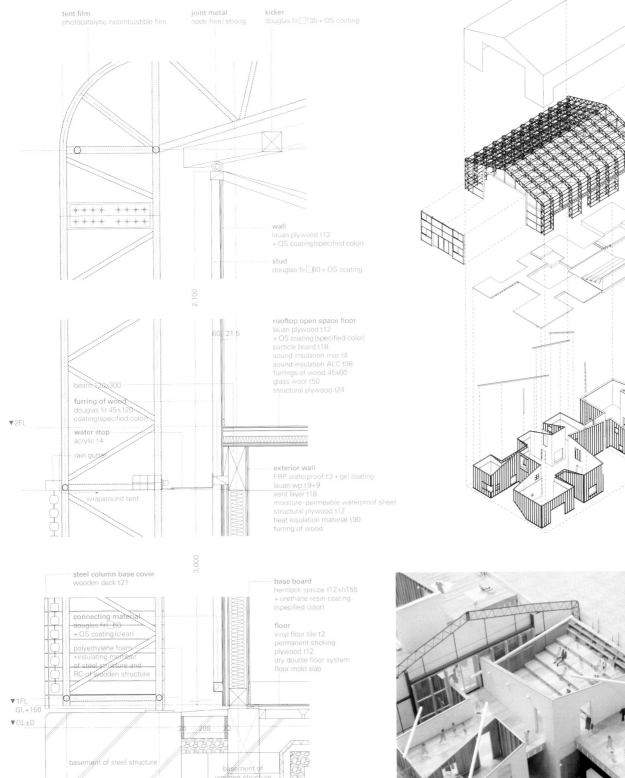

详图1 detail 1

项目名称：Fukumasu Base-Fukumasu Kindergarten Annex / 地点：690-1, Yo-gabashi, Fukumasu, Ichihara city, Chiba, Japan / 建筑设计：Yasutaka Yoshimura 设计参与人员：Hibino Sekkei, Youji no Shiro-supervisor of design / 室内设计、景观设计、工程监理：Yasutaka Yoshimura Architects / 结构工程师：Eisuke Mitsuda / 工程公司：Hirai Construction company / 用途：Kindergarten / 场地面积：3,361.51m² / 建筑面积：512.42m² / 总楼面面积：684.81m² 建筑规模：one story (building A) + two stories (building B) / 结构：steel and wooden structure / 完工时间：2016.3 / 摄影：courtesy of the architect

童戏空间

Five Fields游乐设施_Five Fields Play Structure / Matter Design + FR|SCH Projects
因斯布鲁克大学的游戏屋_Playroom in University of Innsbruck / Studio3 - Institute for Experimental Studies UIBK
格利法达公立幼儿园_Public Nursery in Glyfada / KLab Architecture
瓜斯塔拉幼儿园_Nursery in Guastalla / Mario Cucinella Architects
NUBO_NUBO / PAL Design Group
Niederolang幼儿园_Kindergarten Niederolang / Feld72
ATM幼儿园_ATM Nursery / Hibinosekkei + Youji no Shiro
Chuon Chuon Kim第二幼儿园_Chuon Chuon Kim 2 Kindergarten / KIENTRUC O
Fukumasu基地_Fukumasu Base / Yasutaka Yoshimura Architects
朝日幼儿园_Asahi Kindergarten / Tezuka Architects
希望学校_Wish School / Grupo Garoa Arquitetos Associados
巴兰基亚幼儿园_Baby Gym Barranquilla / El Equipo Mazzanti
埃克苏佩里国际学校_Exupéry International School / 8 A.M.
马萨特兰的玛利娅·蒙特梭利学校_María Montessori Mazatlán School / EPArquitectos + Estudio Macías Peredo
儿童村_Children's Village / Aleph Zero + Rosenbaum
be – MINE游乐场_Play Landscape be - MINE / OMGEVING + Carve

童戏空间_Child's Play / Alison Killing

与那些通常用于指导建成环境设计的价值观相比，面向儿童的设计能让我们前瞻性地接触完全不同的价值观。成年人的需求是多数建筑的设计中心，但在这些面向儿童设计的项目中，成年人的需求则应退居儿童及其学习需求之后。很多学校建筑已经引进了令人惊讶的、富有趣味性的建筑元素；操场上，对能激发想象力的空间的需求也更甚于对功能性元素的需求。本章节探讨了从小学到幼儿园建筑、从露天操场到儿童室内游戏室等多种项目。这些散布在世界各地的项目展示了多种设计方法，其所使用的材料、色彩丰富的外观和建筑的整体空间感不仅激发了孩子们对空间和相关活动的参与，更重要的是，它们营造了一种舒适和富有趣味性的氛围。本章项目向我们展示了另一个与我们平行的世界中专为不同年龄阶段和不同发展阶段的孩子们量身打造的项目。这种建筑对儿童起到了促进的作用，其多种外形和空间能够鼓励孩子们进行社交，并通过娱乐活动和对世界的探索收获更多的知识。

Design for children offers the chance to foreground a different set of values than those which typically guide the design of the built environment. Where most architecture centres the needs of adults, in many of these projects their needs take a back seat to those of the children and their learning. Surprising, playful elements have been introduced to many of the school buildings, while in the playgrounds, functional requirements are less important than space for imagination. This section explores a large number of projects from primary schools to kindergartens and from open-air playgrounds to inidoor playrooms for kids. Scattered around the world, these projects illustrate a large verity of design approaches where materials, colourful shapes and the overall spatiality of the buildings not only stimulate the children's engagement with space and the related activities but, more importantly, they create a comfortable and playful atmosphere. The projects in this section show a parallel world entirely made for kids of various ages and in different stages of their development. The architecture here plays a conducive role, whereby shapes and spaces encourage socialisation, learning through play and a continuous exploration of the world.

童戏空间
Child's Play

Alison Killing

学校建设不仅仅关乎建筑师或那些偶尔灵光一现、认为建筑对于提高学习质量有着重要意义的业主，更是多位影响深远的教育改革家教育理念的重要组成部分。雷焦伊米利亚办学体系是二战后由罗利斯·马拉古奇 (Loris Malaguzzi) 在意大利北部的一个同名地区提出的。该体系认为，包括学校建筑在内的环境是孩子们的第三位教师；教师、其他孩子和环境共同构成了孩子们的教育者。该理念认为这些学校可以促进孩子们的全面发展：它们不仅能提高孩子们的心智技能，还能促进其社交能力和身体素质的发展。教师们不再是孩子们的指导者，而是引领者和合作者。丰富多彩的长期项目可以让孩子自己选择学习的内容和方式；学校建筑和附属的室外空间将为特定类型的学习提供机会。

雷焦伊米利亚类学校有一系列的通用建筑策略，其中之一就是教室应与宽广的外界环境相连。教室的大门应面向供孩子们交流的中央公共区域。教室还要和室外相连，教室内要有足够大的窗户和可以通往操场的大门。学校的中心广场是学生们可以汇集于此、进行互动并学习重要社交技巧的平台。教室内也应配有为大型和小型团体活动准备的活动空间。这类学校的建筑设计要求是经过深思熟虑的，旨在提升学校的教学模式；其中蕴含的许多空间设计理念——将孩子们汇集于此、激发孩子的创造力、让不同的空间可以得到灵活的运用、将学习场所的范围延伸到教室之外、让孩子接触到更多的团体，也都在本章的项目中可以发现。

巴西圣保罗的希望学校 (130页) 虽然没有采用与雷焦伊米利亚学校完全相同的教学法，但也大同小异。Grupo Garoa Arquitetos Associados的建筑师与学校紧密合作，确保其新建筑能够反映并支持这种办学理念。学校旨在强调孩子发展中的多个方面，如身体素质、社交能力、情绪管理、创造能力、能动性和精神成长等，并将这些方面和学生的智力发展置于同等地位。学生参与的活动也由其个人兴趣和能力决定。在被希望学校接手之前，这两个曾经的工业棚屋已经历过多次翻修。在棚屋经过处理并可以遮风避雨后，建筑师们在楼层的布局上就有了很大的自由度，他们选择打造出一种流动的空间序列。在偌大的空间里，每间教室都是独立的空间，很多

With school buildings it is not only architects, or the occasional enlightened client who argues that the building is important for the quality of learning that can take place – it also forms a key part of the philosophies of several influential education reformers. The Reggio Emilia school system was developed by Loris Malaguzzi after World War II in the northern Italian region of the same name. The environment, which specifically included the school building, was considered to be the "third teacher", which, together with the teacher and other children, was seen as the educators of a young child. The idea was that these schools would develop the whole child: not just his intellectual skills, but also his social and physical abilities. Teachers would play the role of guider and collaborator, rather than instructor; multi-faceted, long-term projects would give children an element of control over what and how they learn; the school building and associated outdoor spaces would provide opportunities for specific types of learning to take place.

There are a series of architectural strategies common across the Reggio Emilia schools. One idea is that the classroom should be connected to the wider environment. Classrooms are arranged to open into a central common space where children can meet. They should also be connected to the outside, with classrooms having large windows and doors leading out to the school's grounds. The central piazza of the school allows children to come together, interact and learn important social skills. In the classrooms there are studio spaces and room for large and small group activities. The architectural brief for these schools was carefully thought through to reinforce the teaching model, and many of the spatial ideas – to bring children together, stimulate their creativity, allow flexible use of different spaces and not restrict learning to the classroom, connect children to the wider community can be found in the projects here.

The Wish School (p.130) in Sao Paulo, Brazil is built around a similar, though not identical, pedagogical approach and the architects, Grupo Garoa Arquitetos Associados, worked closely with the school to ensure that their new building would reflect and support this philosophy. The school aims to address the various aspects of a child's development: physical, social, emotional, creative, intuitive and spiritual learning and give them the same weight as

墙体都采用半透明或可移动设计，根据具体活动的需求，教室既可进一步提供封闭的私人空间，也可转变成开放的公共空间。

Yasutaka Yoshimura Architects设计的Fukumasu基地（100页）也采用类似的建筑方法，大面积的棚屋可供内部空间相对自由地排列。建筑师原本想要在原址上改建现有的工业建筑物，但当他们发现这种方法并不合适时，就建造了一栋全新的建筑。他们买下了一个现成的仓库构架，然后在上面铺上一层半透明的白色薄膜，这样就能使整个空间充满漫射的白光。幼儿园一层的设计典雅美观，但相对直白；二楼则有着不规则的空间设计，一些上空空间和小教室与走廊和阶梯式座位区相互连接。整个建筑的形态，以及将诸如内墙的木框架等建筑元素暴露在外的做法，都是为了激发孩子们的想象力。

瓜斯塔拉幼儿园（46页）就建在了雷焦伊米利亚模式的诞生地。它的建筑元素丰富，能与外部空间融合，其内部空间的布局和组织结构以及建材的选择，加之它的建筑颜色、声音效果和触感，能够为儿童提供一场感官之旅。每片区域都通过透明的墙体紧紧地连接在一起，更能激发孩子们的好奇心，带来更愉悦的感受。

巴兰基亚幼儿园（144页）是一家体验导向型的雷焦伊米利亚式幼儿园，孩子们在这里可以学习人际间的交流和相处之道。通透的立面和带有玻璃的圆形教室突出了空间的即时交互性。它暗示着建筑物可以是一个有机体，可以适应此种结合内外空间的新型教学方式。

其他的学校设计则重点关注周围更广阔的环境和社区之间的关系，策略之一则是使所选的建材更贴近当地景观，正如Tezuka Architects为日本南三陆町所设计的朝日幼儿园（120页）一样。他们选用了一些在2011年地震后被海水浸泡致死的巨型树木来建造此建筑的结构。

对于Feld72及其在意大利瓦尔多拉迪索多设计的Niederolang幼儿园（64页）来说，怎样与当地社区联系、联系至什么程度等问题

intellectual education. The activities children take part in are also shaped by their individual interests and abilities. The two industrial sheds that now house the school had already undergone numerous renovations before being taken over by the Wish School. With the shed taking care of keeping out wind and rain, the architects had a lot of freedom in organizing their floor plan and chose to create a fluid sequence of spaces. The classrooms stand as independent objects within the wider space and many of the walls are semi-transparent or movable, offering the possibility of creating closed off, private spaces or open, more public areas, depending on what is needed for any given activity.

Yasutaka Yoshimura Architects' Fukumasu Base (p.100) has a similar architectural approach, with a large shed that allows a relatively free arrangement of spaces inside. The architects had wanted to transform the original industrial building which stood on the site, but when it proved unsuitable, they settled for a newly built one. They bought an off-the-peg warehouse frame, over which they stretched a translucent white membrane, which floods the space with white light. Where the ground floor of the kindergarten has an elegant but relatively straightforward plan, the first floor has an irregular landscape of voids, small rooms connected by walkways and stepped seating areas. The form of the building, as well as the decision to leave structural elements exposed, such as the timber framing of the internal walls, is intended to help stimulate the children's imagination.

A Nursery in Guastalla (p.46) sits in the birthplace of the Regio Emilia approach. Its rich architectural elements – integration with the outdoor space, the shape of the interior spaces, their organization, the choice of materials, the colors, the sounds, the tactile suggestions, create an environment that drives the children's sensory journey. A strong connection between each area is made through transparent walls, further fostering curiosity and pleasure. Children learn the ambiance of dialogue and coexistence in Baby Gym Barranquilla (p.144), an experience-driven Reggio Emilia kindergarten. Permeable facades and circular classrooms with glass highlight the capacity of the spaces exchangeable in time. It is a suggestion of architecture as an organism that adapts to new teaching ways between community inside and outside.

十分重要。他们选择建造了一所带保护性庭院式操场的内敛型学校，运用的是一种基于村庄现存边界和围墙布局的类型学方法。与之对照的是，Hibinosekkei and Youji no Shiro则选择将日本大阪丰中市的ATM幼儿园（76页）建成一座外向型的建筑。从当地住宅区发展起来的社区数量正在萎缩，建筑师希望将这所学校变成当地居民的活动中心，进而重振社区活力，同时也促使幼儿园的孩子们与社区邻里的关系变得更加紧密。这里的许多房间窗户都很大，能让路人看见里面的情况，老师们也能从远处向外监管孩子们在操场上的活动，孩子们也能在教室里观察到其他高年级或低年级的同伴。这种理念就是不仅要在校内建立一种强有力的联系，从而将不同的儿童群体聚集在一起，培养他们的集体感，同时也搭建起学校与外界沟通的桥梁。

ATM幼儿园的外观设计反映了周围建筑的风格，但是仔细观察后，又可以发现一些惊喜之处。幼儿园并未采用当地住宅的瘦长外观设计，但它却搭建了多处阳台，并从立面的不同高度向外凸出，其间还有彼此相连的短楼梯、攀爬设备、架子、长椅和斜坡供孩子们玩耍。还有一个富有趣味性的元素———一张跨越某个院子的大网，其中包含的网状通道可让建筑中的小居民们从结网的屋顶到达下方的庭院。在整个幼儿园内，若要从A地到达B地，会有很多不同的路径可供选择：有的路可直达，有的则具有趣味性；有的简单到连最小的孩子都可以找到，有的却富有挑战性，大一点的孩子才能玩得好。

在布拉德斯科基金会的委托下，圣保罗的Rosenbaum and Aleph Zero建筑师事务所设计的Canuanã寄宿学校的儿童村（188页），是一项由设计者、建造者和使用者共同深度参与的努力成果。特别是儿童们也可以通过一系列工作坊和其他活动参与其中。这些工作坊和活动都因循一项名为"温和改变"的协同运动的原则，Rosenbaum正是这项运动的主要参与者之一。这座面积为 23,344.17m² 的复合建筑被设计成不同的空间，有助于提高在校学生的体验。这个经精心规划的项目还包括学习设施、娱乐区域以及45个6人寝室。值得一提的是，这几个社交区域对孩子们的互动、思想交流和游戏化学习都起着重要的作用。

In other schools the design focus is emphatically on the wider surroundings and the community. Choosing materials that connect closely to the local landscape is one strategy, as at the Asahi Kindergarten (p.120) in Minamisanriku, Japan by Tezuka Architects, who did this through the use of some huge trees killed by salt water after the 2011 earthquake to create the structure of the building.
The question of how far and in what way to connect to the local community was also important for Feld 72 and their Kindergarten Niederolang (p.64) in Valdaora di Sotto, Italy. While they chose to create an inward-looking school with protected courtyard playground, a typology based on the village's existing pattern of boundaries and enclosures, Hibinosekkei and Youji no Shiro chose to look outwards in their ATM Nursery (p.76) in Toyonaka, Osaka, Japan. The communities that had developed in the local housing estates were atrophying and the architects hoped that this building might help to reinvigorate them by becoming a hub for local people, while also connecting the kindergarten's children more closely to their neighbourhoods. Many of the rooms feature large windows, allowing passers-by a view into the centre, giving teachers a view out to supervise playground activities from a distance and meaning that the children get to peek into the classrooms of their older and younger peers. The idea is not only to create strong connections within the school, where the hope is that bringing together different groups of children will foster a sense of community, but also to build bridges to the world outside.
The outward appearance of the kindergarten reflects nearby buildings, but on closer inspection it also has some surprises. The kindergarten doesn't share the long thin form of the local housing blocks, but it does have balconies jutting out at different levels across the facade and connected by short flights of stairs and climbing equipment, monkey bars, benches and slopes to play with. A net spanning over a courtyard is a playful element and includes net tunnels that offer a route from the netted roof to the courtyard below for the building's smaller inhabitants. Throughout the building there are multiple possible routes between A and B, some straightforward, some playful, some easy to navigate even for the youngest children, some more challenging and reserved for the older ones.
Children's Village at the Canuanã boarding School (p.188), commissioned by the Bradesco Foundation and designed by the São Paulo-based office Rosenbaum and Aleph Zero, is the result of a strong participation of all actors involved in the design, building and use of this project. In particular, children have been included in the

雅典格利法达公立幼儿园（34页）的规模（与Canuanã寄宿学校的儿童村）不同，其"城中村"的理念和带有可能发生各种偶遇事件的内部空间的设计是整个项目的核心特征。幼儿园由KLab建筑事务所设计，由四组主要教室建筑和相关设施组成，所有空间都绿植环绕，并面向内部的中心庭院开放。每个包含教室的主要体量都被分为更小一级的体量，以打破大型单色建筑的刻板印象。外部的木制结构和阳台给孩子们提供了更小型的设施，而在各个体量和周围的开放空间之间，还有室外玩耍空间以及树木，这样孩子们就可以在一个充满动感的公共环境中嬉闹了。建筑的内部也采用了类似的空间结构，彩色走廊连接着不同的教室和设施，孩子们可以在整个区域里自由地奔跑。

通过这些学校的设计可以看出，特定类型的建筑和它为之创造空间的教学形式之间有着很强的关联性。Mark Dudek在他有关学校和幼儿园设计的手册中描述了三种主要方法。第一种也许是最传统的：为特定活动设计专门的房间，比如教室、故事角、衣帽间和操场，这类空间的规格和数量通常都由相关规章制度决定。虽然儿童们被视为建筑的主要使用者，但考虑到如安全因素或教师监管的便利性等问题，建筑师通常会优先考虑成人的需求。第二种是学校采用新的教学模式，例如上文提到的雷焦伊米利亚教学法。这种教育理念对建筑环境有着强烈的愿景，要求建筑环境在最大程度上对其给予支持。Dudek认为第三种则是学校的建筑师对建筑的使用者具有强烈的共鸣之心，其学校和操场的设计均以儿童为中心并富有趣味性。

这三种方法在这里所展示的学校项目中都得以反映。希望学校和由EPArquitectos + Estudio Macías Peredo设计的墨西哥玛利娅·蒙特梭利马萨特兰学校（174页）最接近第二种方法，其余的小学和幼儿园则介于第一种和第三种方法之间。所有的学校设计都至少要遵循一定的规定，另外对于很多学校来说，还得满足国家课程的严格标准。虽说第一种是最不引人注目的教学方法，但它是在学校中使用最普遍的教学方法，而且在很多方面都颇有成效。值得一提的是，它本身并不排斥游戏和以儿童为中心的设计。

process through a series of workshops and other activities following the principles of the collaborative movement "A Gente Transforma" of which Rosenbaum is one of the main participants. The different spaces in this 23,344.17m² compound have been designed to be conducive to an enhanced quality of the experience in the school. This project has been designed with an elaborated program that includes learning facilities, recreational areas, and 45 dormitory units for 6 students each. In particular, the several social spaces are important for pupils to interact with each other, exchange ideas and learn through play.

With a different scale, the idea of the urban village, with in-between areas where occasional encounters might happen, is a key feature of the Public Nursery in Glyfada, Athens (p.34). Designed by KLab Architecture, the nursery consists of four main classrooms and related facilities surrounded by green spaces and all face towards a central inner courtyard. Each of the main volumes containing the classrooms is divided into smaller volumes in order to break out the rigidity of large monochromatic building. External timber structures and balconies help provide the kids with falilities on a smaller scale. In between the volumes and around the open spaces there are areas for outdoor play and trees, so pupils can play outdoors in a dynamic and communal environment. A similar spatial structure is replicated inside, where colourful corridors connect the different classrooms and facilities, and kids can run freely in all areas.

With each of these school designs, there are strong links between a particular type of architecture and the form of pedagogy it creates space for. In his design atlas of schools and kindergartens, Mark Dudek described three main approaches. The first is perhaps the most traditional. There are dedicated rooms for specific activities – the classroom, story corner, cloakroom, playground – with the size and number of these spaces often determined by rules or regulations. While children are seen as key users of the building, adult needs are nevertheless often prioritized – for safety, or to be able to easily supervise what is going on, for example. The second type of schools are those with new pedagogical models such as Reggio Emilia described above, where the educational concept also contains a strong vision of the built environment which will best support it. The final model that Dudek identifies is that of architects who have a huge amount of empathy for their users, where their school and playground designs become playful and child centric.

让本文中所列举的许多学校出彩的是它们的娱乐设施——比如8 A.M.设计建成的拉脱维亚Pinki埃克苏佩里国际学校（158页）那穿梭在建筑间隙中的跑道，还有"Spielraeume"项目（20页）中因斯布鲁克大学实验建筑与建筑技术学院第三工作室的学生在他们学院前面的广场上建成的架高平台和等待你去探索发现的隐藏空间。由KIENTRUC O设计的越南胡志明市Chuon Chuon Kim第二幼儿园（86页）则直接从建筑形式入手，将一些小房子一样的空间直接堆叠在一起，并用一部巨大的橙色螺旋楼梯将其连接起来。

砖墙立面以及其上开口模式的设计目的是为了保证自然通风以及看向外部的连续视野。出于类似的原因，教室也是围绕着一个通风的垂直筒形空间而建，每一层都有开向外部的较大洞口。筒形空间的最底层花园蜿蜒走，最顶层则是屋顶露台，拥有俯瞰Saigon河的开阔视野。这种环环相扣的空间安排带来了不断变化的视角，进而能激发住户自身的创造力。

最后一组项目纯粹为了游戏目的而建，这使我们得以从那些已造就大部分建成环境的商业和功能需求中稍事喘息。正如Five Fields游乐设施（12页）的设计者Matter Design and FR|SCH Projects在叙述他们的项目设计方法中所提到的那样，"专为孩子们设计游乐设施的任务既独特又富有挑战性。较之成年人的日常生活中都必有规范（项目）通达性和功能性的标准，游戏既算不上是什么标准，更算不上严格意义上的功能性。"但游乐场的设计还是需要遵守一些原则，比如物体的尺寸一定要适合儿童，也要允许游戏存在一定的风险，以使它有足够的挑战性，进而使孩子们感兴趣。教育改革家约翰·杜威（John Dewey）将游戏视为孩子们在其中进行学习的一条关键途径，但他也强调游戏本身也必须要合格，游戏活动就是要提供快乐，要让孩子们愿意参与，参与的目的纯粹就是为了游戏，而不需要考虑特定的学习成果。在实践过程中，这一点就意味着好的项目设计要在为数不多的功能要求前提下付出大量的想象力，这不仅是为了年幼的使用者，同样也是为了建筑师自身。

The buildings shown here cover this spectrum. The Wish School and the María Montessori Mazatlán School in Mexico by EPArquitectos + Estudio Macías Peredo (p.174) belong most closely to the second group. The other primary schools and kindergartens tend to sit somewhere between the first and third categories. All school buildings must conform to at least some regulations and for many schools, also to the strictures of national curricula. While the first category is the least immediately compelling, that teaching model is common and, in many ways, effective. It should also be noted that it does not, of itself, rule out all playfulness and child-centred design.

It is the playful elements that stand out in many of the schools here – the running track that passes through a gap in the building at the Exupéry International School in Pinki, Latvia by 8 A.M. (p.158), or the raised platforms and hidden spaces waiting to be discovered in the Spielraeume (p.20) built by the students of studio 3 in the Department of Experimental Architecture at the University of Innsbruck, on the square in front of their faculty. At the Chuon Chuon Kim 2 Kindergarten in Ho Chi Minh City, Vietnam by KIENTRUC O (p.86), this starts with the building's form – a pile of tiny house shapes stacking on top of each other, with a giant orange spiral staircase to connect them all.

The brick facade, with its pattern of openings, is designed to allow natural ventilation as well as continuous views out. For similar reasons, the classrooms are arranged around an airy vertical core, with large openings to the outside at each level, starting at the ground floor where the garden flows under the building and ending with a roof terrace with an expansive view over the Saigon river. This interlocking series of spaces, with its ever-changing views is intended to be stimulating for the users and to inspire them to be creative themselves.

The last group of projects here are given over entirely to play, an activity which allows a step away from the commercial and functional demands that shape much of the built environment. As Matter Design and FR|SCH Projects say in describing their approach to designing the Five Fields Play Structure (p.12), "designing a play structure intended for kids is a unique and challenging project. While there are irreducible standards that manage accessibility and function in the daily lives of adults, play is neither standard nor strictly functional." There are still some requirements that guide playground design, such as dimensioning objects to make them suitable for children, and for play to allow an element of risk so that it's challenging enough to be interesting to children. Education re-

Five Fields游乐设施的结构形式呈线性，第一眼看上去，它的门、楼梯和走廊的设计都相对直白，但随后你就会发现其实它并非如此。通道根本不会通往任何一个地方，或者在半路上就突然终止了。想要到达同一个平台会有多种路径，如爬梯子、走楼梯，甚至是爬墙过去，只不过有的途径比另外一些更具挑战性。在保持趣味性的同时，它还会为孩子们带来成就感，即使是随着孩子们的能力与日俱增，也同样如此。澳大利亚悉尼的NUBO幼儿园（54页）则是一个具有激励性和包容感的游戏中心，这里鼓励孩子们去学习，去探索，并激发他们无尽的想象力。为了满足孩子们的好奇心并突出"纯粹娱乐"的理念，NUBO设置了不同的可供孩子们安全探索和使用的学习空间并引以为豪。

当地景观是OMGEVING + Carve获奖设计的出发点。他们设计的be-MINE游乐景观（206页）位于比利时的贝林根，这项工程是前煤矿大型改造项目的一部分，景观就建在矿渣堆（或称"terril"）上。建筑师在碎石山的一侧打造了一处有着凹凸棱格的水泥斜坡，还设置了一部任何山上游乐场都不会错过的巨型滑梯，另外还有一片可供攀爬的木桩森林。随着高度的不断上升，攀爬的难度也在逐渐增加，由此可使不同年龄和不同能力的孩子们都参与其中，进而鼓励孩子们通力合作，共攀高峰。

在这些项目中的游戏提供了前瞻完全不同价值观的机会，而非那些在通常状况下驱动建成环境设计的监管和商业思维。

建筑物和游戏场地的设计仍需满足一些功能性的条件，但成人的需求将不再被优先考虑，设施、楼梯和通道的尺寸也将被缩小到与儿童相适应的理想尺寸，并经常会达到成人无法使用的程度。在提供一套好设计构成新标准的同时，童戏空间不仅为孩子们提供了玩耍的地方，也为创造这些设计的建筑师们提供了一个全新的、可供选择的视角来观察建筑最终究竟可以成为什么样子。

former John Dewey saw play as a key way in which children learn and yet also emphasized that to qualify as play, an activity that should be done for pleasure, enjoyable to take part in, and done purely for its own sake, not with specific learning outcomes in mind. In practice, that tends to mean that design has to strike a balance between a small number of functional requirements and a lot of imagination, both for the child user and in the best projects, on the part of the architect too.

Five Fields Play Structure with its rectilinear forms – doors, stairs and corridors – appears relatively straightforward at first glance, but then reveals itself to be anything but that. Passages lead nowhere, or end suddenly in mid-air. Multiple routes – ladders, stairs, climbing walls – lead to the same platform, some more challenging than others. It allows a sense of progression, with the play structure remaining interesting, even as children grow in ability.

NUBO (p.54) in Sydney, Australia, is a stimulating and inclusive play centre that encourages learning, exploration, boundless imagination. To satiate kids' curiosity and emphasize the concept of "Pure Play", NUBO boasts a variety of learning spaces to be safely explored and enjoyed.

The local landscape was the jumping-off point for the design of OMGEVING + Carve's competition-winning design. Their Play Landscape be-MINE(p.206) in Beringen, Belgium is part of the wider redevelopment of a former coal mining site and is built on top of the mine's slag heap, or "terril". The architects have created a tessellated concrete slope on one side of the terril, as well as a giant slide that no mountain-based playground could be without and a forest of poles to clamber up. The climb gets more difficult as you ascend, so it can accommodate children of different ages and skills levels, encouraging them to work together to make it to the summit.

Play in these projects offers the chance to foreground completely different sets of values than the regulatory and commercial concerns which typically drive the design of the built environment.

The buildings and playgrounds still have to meet some functional requirements, but the needs of adults are deprioritized and the sizes of furniture, stairs, tunnels, reduced to the ideal size for children, often to the exclusion of adults. In bringing a new set of criteria for what constitutes a good design, it offers not only the children, but also the architects creating these designs, space to play in and gives a refreshing, alternative view on what architecture can be.

朝日幼儿园
Asahi Kindergarten

Tezuka Architects

最初的朝日幼儿园在2011年3月11日日本东北部大地震中毁于一旦。在联合国儿童基金会日本委员会的资助下，Tezuka建筑师事务所对朝日幼儿园进行了设计，并以2011年海啸中被海水浸泡过的树木为建材，将其重建于山丘之上。对当地村民来说，这些树木具有特殊的象征意义，因为之前它们曾被栽种在通往大雄寺的步道上。大雄寺的主殿在山中高处，刚好可以使它在过去发生的海啸中得以幸存。寺内的住持曾让村民们来庙中避难，因此很多村民都在这场海啸中幸存了下来。项目旨在表达树木不仅可作为幼儿园的建材，同时也是村民们的精神家园。利用枯木为Tohoku的下一代建造一座新的幼儿园，这使每一位村民，甚至是每一位日本人都重拾希望。建筑的每一部分，包括结构、地板和扶手都来源于这些1611年海啸过后栽种的树木，到2011年本次海啸，正好400年。项目采用了传统的细木作及没有任何金属连接件的楔子，这些老手艺已使日本传统建筑存活了1300多年。建筑中有一根截面尺寸为600mm×600mm的巨柱，一如它当年生长时那样矗立在地面上。这为或许会在此后400年中再次遭受海啸侵袭的孩子们传达着一种信念。

海啸已经过去多年，已有越来越多的孩子回到了南三陆町，一期幼儿园的规模已无法容纳更多的儿童，因此幼儿园决定马上启动二期工程建设。

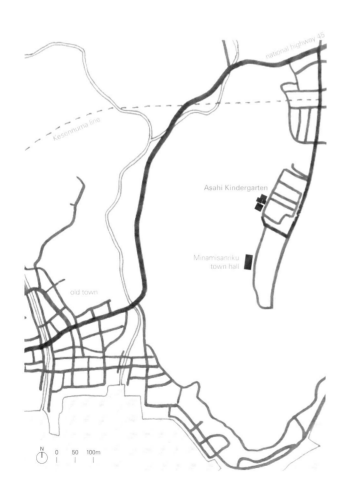

不幸的是，场地周遭的环境已经发生了完全的变化。随着灾后重建工作的开展，大规模的开发工程也匆忙上马，这里原有的郁郁葱葱的景色不复存在，山体也遭到了严重破坏，只有周边的花园还保持了原有的高度。二期工程更像是在山坡上建了一座带有宝塔的综合性寺庙，因为人们认为大雄寺就像是幼儿园的母亲一样。现在这里有三座加建建筑，每座都屋檐深深，并由一部长长的楼梯连接在一起，一期建造的公园位于山顶位置。

曾栽种在大雄寺参拜步道上的树木已使用殆尽，因此无法再制作巨柱，设计师决定采用更经济的方式，利用尺寸合理的材料进行建造。

然而，幼儿园承载的信念必须要传达给400年后的孩子们，因此二期建筑所有的接口设计仍采取一期的传统联锁方法，以确保这座建筑能够久久屹立不倒。

The original Asahi Kindergarten was lost in the Tohoku earthquake on 11 March 2011. Funded by the Japan Committee for UNICEF, Tezuka Architects designed and reconstructed the Asahi Kindergarten on a hill by using the dead trees by the salt water of 2011 tsunami. These trees are meaningful symbols for the local villagers as they were planted along the approach to Daioji Temple. The main temple on the hill is just high enough to elude tsunamis in the long history. Many villagers survived because the priest of the temple used to teach them to escape to the temple. The aim of the project was to express that the tree was not only the building material for the kindergarten but also a

home to the spirit of the villagers. Reusing the dead trees to build a new kindergarten for the next generation of the town reaffirms hope held by everyone in Tohoku or Japan. Every piece of the building, including structure, floor and handrail, was carved out from these trees which were planted after the tsunami in 1611, exactly 400 years before the tsunami in 2011. Traditional joinery and wedges without any metal joints were used, because these old techniques have made Japanese traditional architecture survive more than 1,300 years. There is a massive column with sectional dimension of 600mm x 600mm erected on the building as how it originally stood on the ground. The project bears a message for those children who will likely encounter a tsunami in the next 400 years.

Several years have already passed since the earthquake, more children have returned to Minamisanriku Town. The first phase building became insufficient and the kindergarten decided to carry out the second phase of construction in a hurry.

Unfortunately, the area around the site is now completely transformed. There was hasty large-scale development accompanying earthquake recovery. The lush landscape has gone. The hill is greatly scraped; only the surrounded garden area remains its original altitude. The site became a landscape like a castle tower built in a flat residential area. The second-phase construction is very much like a temple complex with a pagoda on the slope of the mountain. It is the result of being conscious of the Daioji temple which is the mother of this kindergarten. There are now three additional buildings with deep eaves, and a long staircase is connecting these buildings together. The garden of the first phase building stays on top of the hill.

Since the big promenade trees have been used up, large sectional column cannot be made anymore. It was decided to construct economically with rationally sized materials. However, the message to children after 400 years has to be kept. All joints are still designed as traditional interlocking like the first phase so it stands over a long period of time.

项目名称：Asahi Kindergarten Phase I & Phase II / 地点：Minamisanriku, Miyagi Prefecture, Tohoku, Japan / 设计师：Tezuka Architects
主创设计师：Takaharu Tezuka, Yui Tezuka / 结构工程师：TIS & Partners / 业主：Phase I - UNICEF, Phase II - Asahi Kindergarten
场地面积：Phase I - 1,819.69m², Phase II - 3,300.12m² / 建筑面积：Phase I - 389.48m², Phase II - 359.48m² / 总楼面面积：Phase I - 287.64m², Phase II - 280.72m²
材料：timber structure / 设计时间：phase I - 2011.9 ~ 2012.2, phase II - 2014.12 ~ 2015.9 / 施工时间：phase I - 2012.2~7, phase II - 2015.9~2016.12
摄影：©Kida Katsuhisa / FOTOTECA (courtesy of the architect)

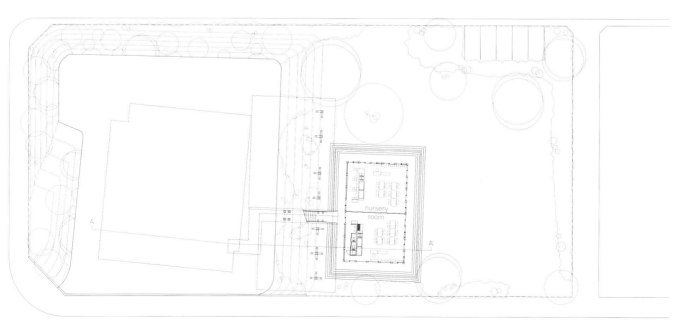

一层 ground floor

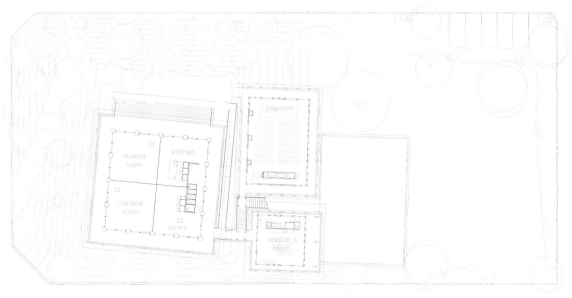

二层 first floor

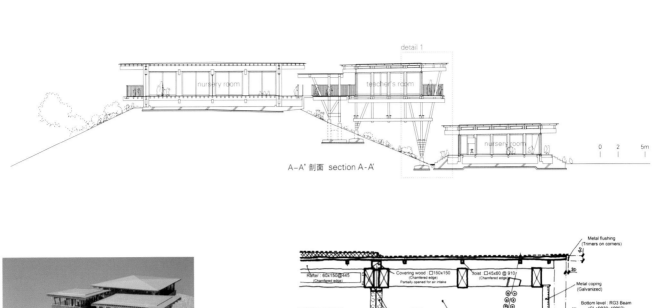

A-A'剖面 section A-A'

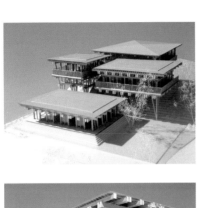

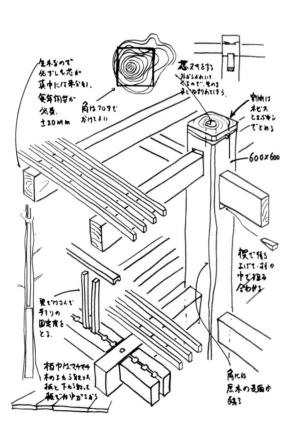

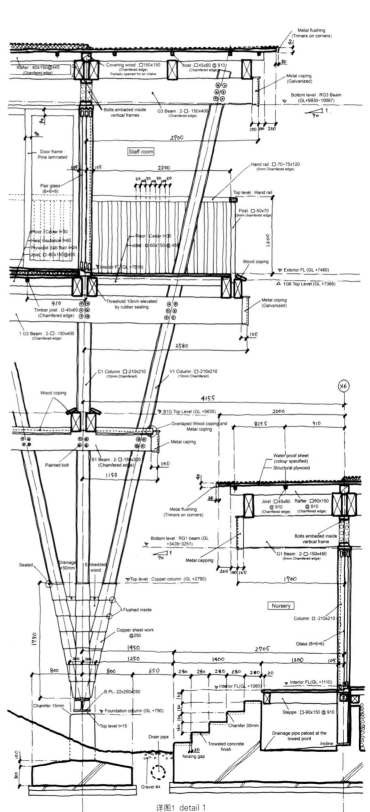

详图1 detail 1

希望学校
Wish School

Grupo Garoa Arquitetos Associados

教室并非是一个容器,而是学生们发展的支点,这是希望学校的建校理念。其教育法植根于这样的理念,即将学生视为独立的个体,并认为他们的身体、情绪、社交、文化、形体、创造力、直觉和精神等层面的发展和理性智慧层面的发展同等重要。学校不仅提供教学内容,还会根据每一个孩子的梦想和潜力重构和强化孩子们的学习过程。

项目旨在建造一座能反映出学校办学理念的建筑。Caio Vassão是一名建筑师,同时也是一名城市规划师,他与其他建筑师、学生和教育工作者通力合作,刻苦钻研教育方法中所包含的复杂互动关系,以确保最终的建筑可以全面实现这些动态关系。由此产生的结构不仅全方位展示了待解决的实践性和功能性问题,还反映了人们对其或抽象或直接的感官期望。

建筑原址面积为15mx50m,其上有两个为满足不同需求已经过数次翻新和改造的相连棚屋,为了把这里重建成一个孩子们可以通过共处和静坐进行学习的场所,需要对整个结构进行彻底的干预改变。平面的布局方法是在楼层中设计多个可扩展或可收缩的小分区。尽管建筑结构将这些小分区的边界也包含在内,但边界的设计却十分轻巧,足够鼓励孩子们更进一步,发挥自己的想象力,像摆弄灵活的物件一样,对小分区进行重新设计,进而实现整个大空间的变化。走廊的唯一作用就是让人们穿梭其中,但值得一提的是该校却没有走廊。本项目中,建筑的每一部分都不仅是一间正式的教室,它们都能让学生们吸收到知识。之所以去除走廊,也是为了表明这一观点:学生们应自由选择路径,选择要去探索什么,要成为什么样的人,并通过不同的方式与世界进行互动。

学校与周围地块相邻并排成一排,因此需要在楼顶板上开设天窗或洞口,为一楼提供自然光。通往最上层的楼梯安置在这些高层天花板上的开洞下方。隔断墙由不同的材料制成,起到过滤光线的作用,使每个分区都能得到各自所需的光照量。每面隔断墙旁都对应设置有一个架子,作为每个房间的隐私保证措施。

建筑的主体侧面部分由两种类型的房间构成:一种是带有固定半透明封闭设施的房间,另一种是以活动面板来界定空间的房间。如果在周围举办了一些活动,面板就可以被利用起来,可以将其当成壁橱、书架、衣架或乐器架等。向周围移动这些面板还可以改变空间的结构,将教室延展到周围的区域。这样就能在更多的活动中实现更好的互动效果。

教室有点像是整个连续空间中的支持区域,是孩子们进入学校后的庇护所。不同于一般的规规矩矩的设计,建筑灵活的边缘设计从两端展开,其内部空间既离散又统一,既拒绝清晰的边界设定,也没有固定的功能,无论是正式还是非正式的学习空间,无论是哪个房间还是其多变的相邻空间,目的都是为了满足开放式教学环境的不同需求。

通过在学校进行视频拍摄,对学生和教师在不同活动中对各种空间的使用情况进行了后期调查。在视频中我们能看见,讲座在室外举行;孩子们聚在一间教室内进行小组研究;有一个孩子在长椅上下象棋;一个女孩躲在楼梯下读书;老师们正在餐厅的桌子前开会。这项分析证实了此项目使命的完成:打造一个能促使学生们重新认识学习过程的建筑,并使学生们置身于一个能充分认识集体生活固有矛盾的空间之中。

Wish School is built on the idea that the classroom is not a container, but a point of support for the students. Its pedagogy is rooted in the idea of helping students as individuals, with the physical, emotional, social, cultural, corporeal, creative, intuitive, and spiritual aspects of a child considered just as important as rational intellect. The school goes beyond providing educational contents, taking each child's dreams and potential into account to reframe and strengthen the child's learning process.

The goal of this project was to create a building that reflected the school's philosophy. Architect and urbanist Caio Vassão, in collaboration with other architects, students, and educators, delved heavily into the complex interactions involved in this teaching approach to make certain these dynamics were fully realized in the final product. The resulting structure is not only a panorama of practical and functional questions to be addressed, but also of sensory expectations both abstract and literal.

The original site (15x50m) contained two connected sheds that had already been renovated and repurposed many times for multiple purposes. The structure required thorough intervention in order to be recreated as a place where students could learn through coexistence and meditation. The floor plan was approached as a space containing both expansive and contractive zones. Although the structure would contain the borders, they would be tenuous enough so that the students would be encouraged to go beyond and use their imagination to redefine the spaces as active subjects, and then transform the space. Notable is the absence of hallways, whose sole purpose is as a venue of movement. In this project, every part of the architecture goes beyond the role of a formal classroom and serves to help students assimilate knowledge. The hallway is removed to support the idea that students are free to choose their own paths, what to explore and what to let be, and interact with the world in different ways.

The building is aligned adjacent to the neighbouring lot, which required zenith openings and cuts in the slabs to provide natural lighting to the ground floor. The stairs that lead to the upper floor are positioned below these open-

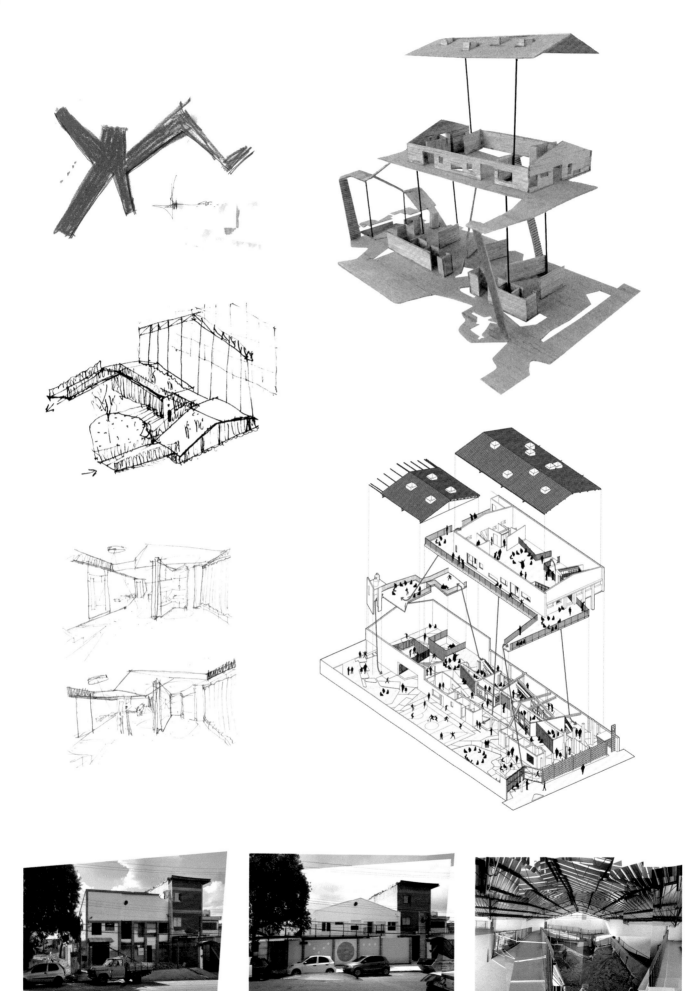

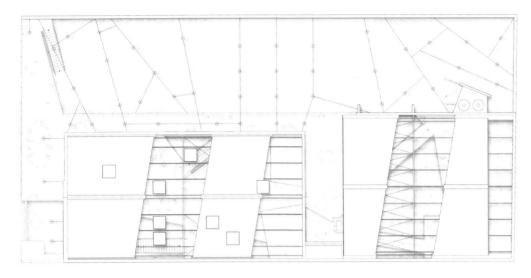

屋顶 roof

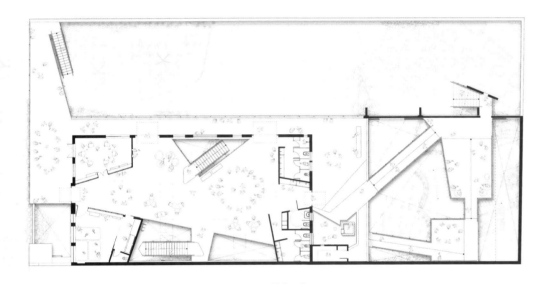

二层 first floor

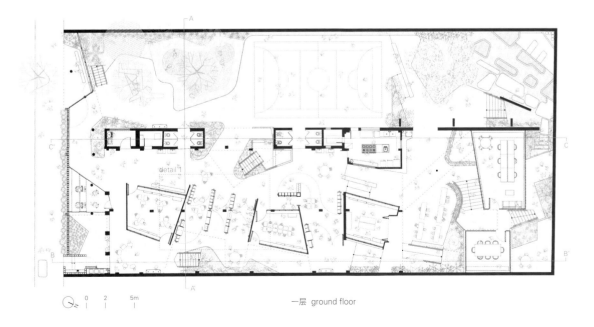

一层 ground floor

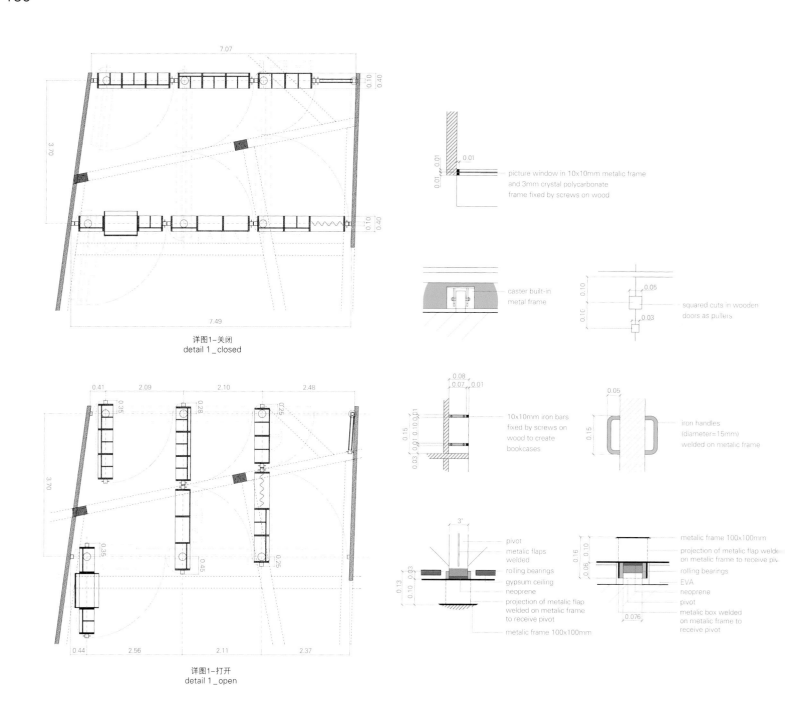

详图1-关闭
detail 1_closed

详图1-打开
detail 1_open

ings in the high ceilings. Different materials are used for the partitions that serve as light filters, allowing each program to receive the quantity of light it requires. The shelves serve as a counterpoint to the partitions, giving each room a measure of privacy for its functions.

The main edges are composed of two types of rooms: rooms with fixed and translucent seals, and rooms with movable panels that delimit the space. The panels function as support for the activities taking place around them, like closets, bookshelves, backpack hangers, instrument shelves, and more. Moving the panels around changes the configuration of the space, allowing classrooms to be expanded into surrounding areas. This allows more activities more interaction.

The classrooms are a sort of support area for the continuous spaces, acting as a sanctuary for the students as they enter the school. Boasting non-orthogonal design, the structure's dynamic edges unfold at the ends, and its spaces are both discrete and unified, refusing to be defined by clear boundaries and delimited functions. The formal and informal learning spaces, and the rooms and their varied contiguous spaces, aim to meet the diverse demands of open pedagogical environments.

A post-development survey was conducted through a video shot to analyze the appropriation of the different spaces by students and educators for different activities. Lectures took place outside rooms; children conducted group research in one of the rooms; a child was playing chess on a bench; a girl was hiding under the stairs reading a book; the teachers were having a meeting at the cafeteria tables. The analysis validated the project's mission to create a structure that serves as a catalyst for re-appreciation of learning and as a space that exposes students to the inherent conflicts of collective living.

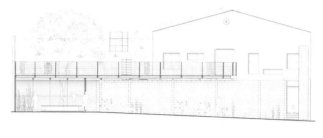
南立面 south elevation

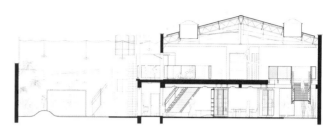
A-A' 剖面 section A-A'

B-B' 剖面 section B-B'

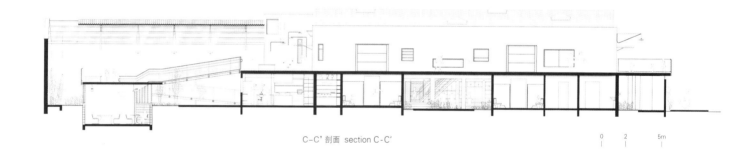
C-C' 剖面 section C-C'

0 2 5m

项目名称：Wish School
地点：Rua São Gil, 159. Tatuapé, São Paulo, Brazil
建筑师：Alexandre Gervásio, Erico Botteselli, Lucas Thomé, Pedro De Bona_Garoa
中介：Caio Vassão
合作者：Nathália Lorena, Liene Baptista, Thaís Coelho, Vinícius Costa
出版合作：Vitor Pissaia
结构工程师：Telecki Arquitetura De Projetos, Carolina Milani De Oliveira
景观建筑师：Tristan Bonzon
电气工程：HV Engenharia
液压顾问：ENG11
施工管理：Grupo Garoa Arquitetos Associados
承建商：Camargo Rosa Empreiteira
木作：MKT 509
锁具：Colombo Esquadrias
电工：Forsaitt
数据运行：Express Technology
场地面积：1,275m²
建筑面积：1,166m²
完工时间：2016.7
摄影：©Pedro Napolitano Prata (courtesy of the architect)

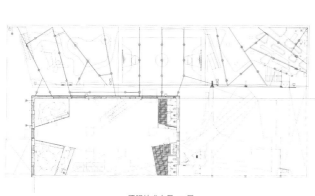

照明技术布局-二层
lighting technology plan_second floor

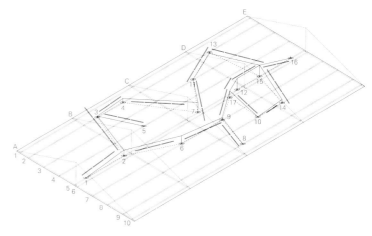

照明技术轴测图
lighting technology axonometric

巴兰基亚幼儿园
Baby Gym Barranquilla

El Equipo Mazzanti

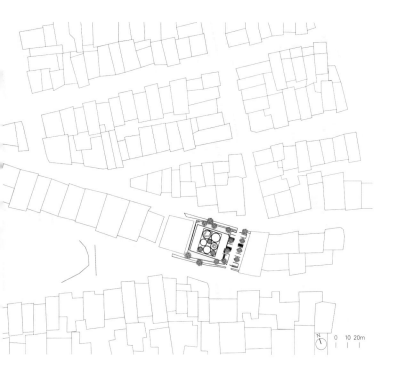

该项目体现了一种综合性理念,有助于实施一套全然不同的教育体系。在这里,空间成为引导学生进行学习的第三任教师。项目中很多形式、概念和技术方面的决策都以强调儿童体验式教学的雷焦伊米利亚教育体系为基础,造就了这所体验导向式的幼儿园。幼儿园强调的体验之一就是幼儿园与城市和社区民俗文化之间的联系。现在,老师和学生们组织的各种活动都试着把幼儿园内的活动与当地居民联系起来。建筑物的外围采用可透视设计,可使孩子们和园外的路人看到彼此的情况。巴兰基亚是哥伦比亚的一座沿海城市,气候温暖,这种立面设计可以实现自然通风,保障孩子们的舒适度。幼儿园总的理念是提供一处对话、共处和能进行高效率学习的场所;这种建筑模式应重视空间作为学习和社交场所的整体性。从这个意义上来说,教室、交通流线区域、露台和外部空间的设计都将在每个机构教育过程中都存在的典型活动和事件中呈现出多样性和复杂性。

El Equipo Mazzanti希望能鼓励园方在园中每个空间里都尽可能多地组织各种活动。圆形教室的设计与孩子们所参加的工作坊相呼应,工作坊以人类感觉与大自然为主题。有时候,园中一些活动空间的安排显得有悖传统,比如厨房居然和音乐室相邻,摄影工作室甚至位于园艺区旁。因为这里房间的"悬浮"设计能够尽可能地保证中间区域的可透视性和清晰性,因此每间教室里的孩子都可以看到阳台上或别的工作坊中其他孩子的情况,从而让孩子们不仅能同步了解自己是如何被展现在别人的眼中的,同时也能了解同学们的情况,处理人际关系,增进自然知识。教室的圆形形状为不同目的的活动提供了灵活的空间,展现了空间在时间上的可交换性。它们之间的空隙还形成了一楼的露台和二楼的"洞口",这些都为儿童们的学习创造了条件。最后,一向被传统观点认为能起控制作用的走廊也打破了常规,变成了充满趣味的场所,不仅能对孩子们起到激励作用,也能让孩子们更快地融入社会环境。

孩子们的童年中有四分之三的时间都将在幼儿园中度过。因此,

每个元素都要从儿童的尺度出发，把建筑营造成一个亲切而又舒适的场所，让孩子们在这种教育环境中感到舒适。巴兰基亚幼儿园是教育建筑中的典范之作，它应被视为一个有机体，它能够做出改变，并适应适用于孩子、教师、家庭和整个社区的新型教学方法。

The project presents a complex that contributes to implementing a different pedagogical system easily, in which space acts as a third teacher, as a guide for learning. Based on Reggio Emilia pedagogical system, which emphasizes children's learning through experiences, many of the formal, conceptual and technical decisions made this an experience-driven kindergarten. One of the highlighted experiences was the connection with the folkloric culture of the city and the neighborhood. Today, children and teachers make activities trying to relate locals to what they are doing inside. The permeable perimeter of the building lets children see what happens outside and people see what happens inside. As Barranquilla is a coastal city in Colombia with a very warm climate, this type of facade also allows natural ventilation to take place, guaranteeing comfort for children.

The general concept is to produce an ambiance of dialogue, coexistence and quality learning; the architectonical model should value the totality of the space as a place of learning and social relationships. In this sense, the classrooms, circulation areas, patios and exterior spaces are designed to achieve diversity and complexity from typical activities and events of the educational process of each institution. El Equipo Mazzanti wanted to encourage the most activities possible in every space of the kindergarten. The circular classrooms respond to the different workshops the children attend to, which are themed by human senses and nature. Spaces with activities that sometimes sound contradictory to the traditional buildings are arranged together, like the kitchen next to the music room, or the photography studio next to the gardening area. Because the rooms inside "float" to allow the central area to be as clear and permeable as possible, children in each classroom can see what others are doing on the patios and in other workshops, allowing them to learn in a horizontal way not only about how they are being shown but about their classmates, relationships, and nature. The circular shape of the classrooms propitiates flexible spaces for different purposes, showing the capacity of the spaces to be exchangeable in time. They also create voids to form patios on the ground floor and "holes" on the first floor, which creates conditions that embrace learning. Finally, the classical idea of a corridor as a mechanism of control is transgressed and changed to ludic spaces of stimulation and social integration.

A kindergarten is a space where kids spend three-quarters of their childhood. Thus each element should answer to kids' dimensions to turn the building into a cozy and friendly place where children feel comfortable in the educational environment. Baby Gym Barranquilla is an exemplary educational architecture that should be thought of as an organism able to change and adapt to new teaching ways among kids, teachers, families and all the community.

©Courtesy of the architect

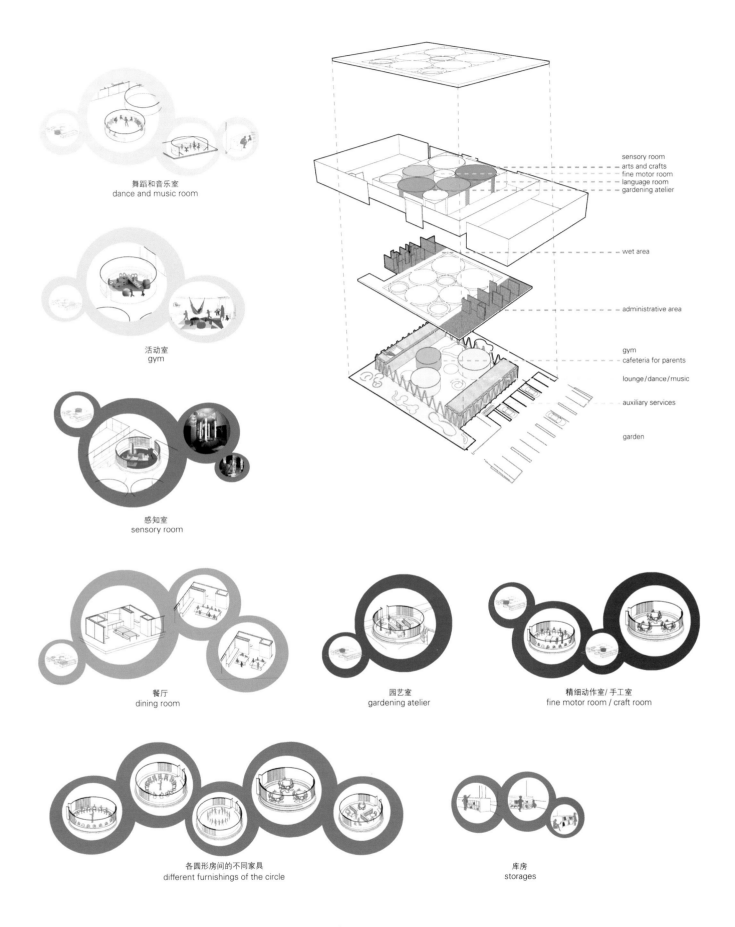

舞蹈和音乐室
dance and music room

活动室
gym

感知室
sensory room

餐厅
dining room

园艺室
gardening atelier

精细动作室/手工室
fine motor room / craft room

各圆形房间的不同家具
different furnishings of the circle

库房
storages

sensory room
arts and crafts
fine motor room
language room
gardening atelier

wet area

administrative area

gym
cafeteria for parents
lounge/dance/music
auxiliary services

garden

项目名称：Baby Gym Barranquilla / 地点：Barranquilla, Colombia / 建筑设计：Giancarlo Mazzanti - El Equipo Mazzanti
项目团队：Juan Manuel Gil, Liv Johana Zea - architects; Felipe Vejarano, Paula Santander, Joana Peixoto, Juliana Alzate, Jessica Jaramillo,
Andres Correa - interns / 业主：private / 建筑面积：1,100m² / 饰面材料：concrete, U-glass / 设计时间：2011.1—7 / 施工时间：2011.9—2012.8
完工时间：2016.1 / 摄影：©Rodrigo Dávila (courtesy of the architect) (except as noted)

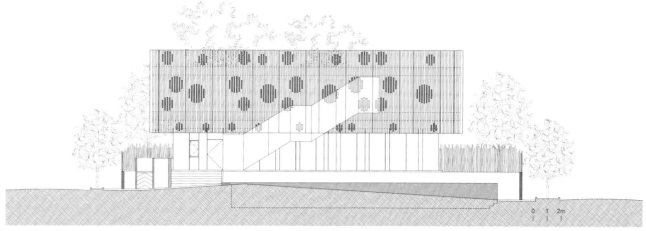

东南立面
south-east elevation

西南立面
south-west elevation

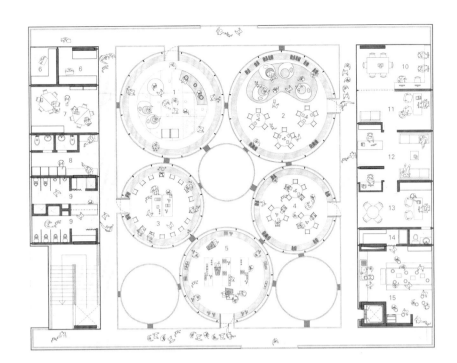

1.感知室
2.手工室
3.精细动作室
4.语言室
5.园艺室
6.存储室
7.辅助室
8.教师更衣室
9.卫生间
10.教师办公室
11.班主任室
12.接待室
13.主任办公室
14.库房
15.餐厅

1. sensory room
2. craft room
3. fine motor room
4. language room
5. gardening atelier
6. cellar
7. auxiliary classroom
8. teachers' dressing room
9. wc
10. teachers' room
11. coordinator's office
12. reception
13. director's office
14. storage
15. dining room

二层 first floor

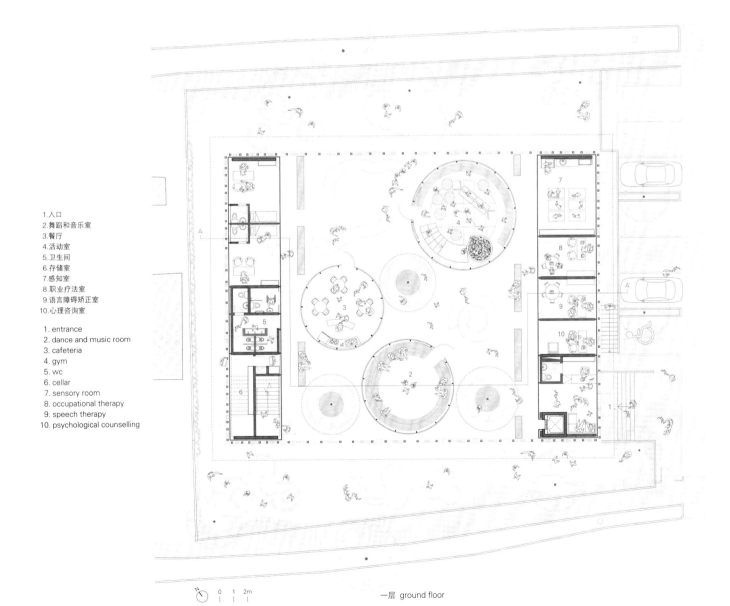

1.入口
2.舞蹈和音乐室
3.餐厅
4.活动室
5.卫生间
6.存储室
7.感知室
8.职业疗法室
9.语言障碍矫正室
10.心理咨询室

1. entrance
2. dance and music room
3. cafeteria
4. gym
5. wc
6. cellar
7. sensory room
8. occupational therapy
9. speech therapy
10. psychological counselling

一层 ground floor

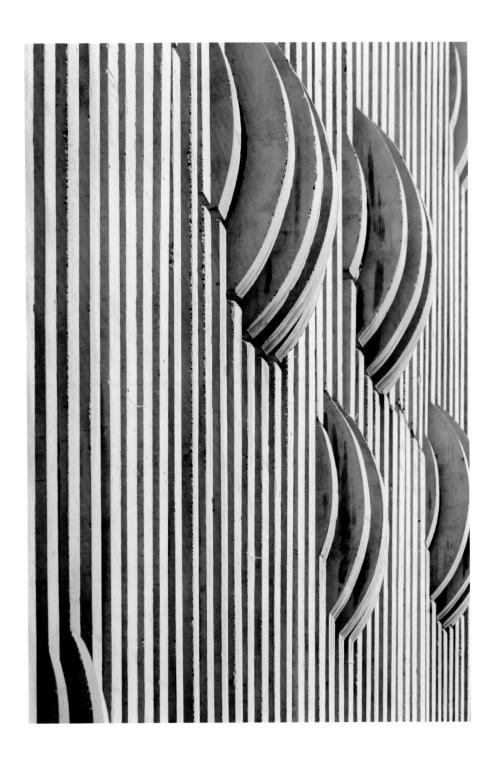

A-A' 剖面 section A-A'

埃克苏佩里国际学校
Exupéry International School

8 A.M.

西立面 west elevation

南立面 south elevation

东立面 east elevation

北立面 north elevation

2016年秋，一家新的教育机构在Pinki开始运行了。它是一所包含了学校和幼儿园的综合设施，遵循一种以儿童为中心并鼓励创新的新型教育模式，这一点从其建筑中也能得以体现。

幼儿园班级按年龄段分为四组，这些班级按照年龄从小到大的顺序依次排列在圆形教学楼中，最大班的学生还进行小学阶段的教育。每个年龄组的儿童都会学习多部世界文学作品，以加深他们对周围世界的了解。上小学之前，他们会通过书本体验到森林、村庄、城市、世界和宇宙等不同的环境。建筑内部和院子的功能反映了他们所探索的文学作品的主题，会让他们亲身接触到宇宙的构成要素，如苔藓、树木、牧场、草坪、院子和沙子。

幼儿园的二层为教室和走廊，它们在低年龄段和高年龄段的孩子们之间提供了一种联系感：在二楼的走廊可以看见低年龄段孩子们的操场，而低年龄段的孩子们也能看到高年龄段孩子们的班级活动。学校建筑的一层专门用于学校生活的公共活动，包含大厅、运动区和孩子们的衣柜；二楼则配有图书室，还有为语言课、艺术与手工课、计算机课、化学课和物理课配备的专用教室。建筑的两部分由"梦之桥"连接。学校这一侧有自己的入口，立面上拥有大面积的开洞，开洞处还提供了有顶的户外区域。一条100m长的跑道贯穿其中，供学生上体育课

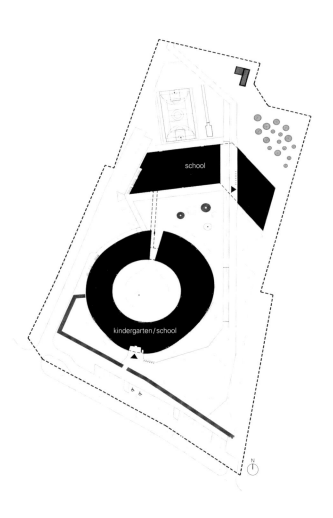

使用。因为地面条件和空间结构的原因，设计师为建筑选用了整体式钢筋混凝土结构。此种方法的优点在于可在建筑物的外围沿着太阳的运动轨迹设置巨大的屋顶，对太阳光进行遮挡，从而使教室免受阳光直射和高温带来的影响。

垂直排布的铝质立面也起到了类似的作用，允许太阳光在短时间内直射穿过立面上的玻璃。二楼教室内的大窗朝向东北方。教室内还安装了天窗，以保证白天的采光效果。

学校的立面上安装了狭长的窗户，这样可使更多的日光照射进来，为孩子们带来更多的温暖，同时也凸显出部分封闭式百叶窗的设计美感。

立面选用的建材具有很长的使用寿命，且短时间内无需翻修。其上木制装饰的原料是西伯利亚落叶松，无论是在适应当地气候的能力方面，还是在可持续性方面，它都属最佳选择。大楼没有采用中央暖通空调设备，而是采用了一套建筑管理系统来进行监测，可对用户控制和外部天气状况做出响应，实现幼儿园和学校区域内所有房间的高质量空气流通。通风口和微孔织物材料结合使用，织物可使空气得以缓慢而持续地流通，且易于维护和清理。噪音较大的房间（如幼儿园的分组区域和教室）则采用了隔音板作为其吊顶或墙壁装饰的材料，可以提升使用者的舒适度。

桦木可以反映当地人的精神，由它制成的胶合板不失为室内建材的一种选择。天然油毡和粉笔漆也因为它们在儿童的教育发展、健康和房间的生态友好性方面所拥有的可持续性而得以入选，毕竟在接下来的几年里，学生们会在这些房间中度过他们的大部分时光。

In autumn of 2016, a new educational establishment opened doors in Pinki. Uniting school and kindergarten in one complex, it follows a new child-centered educational model that encourages creativity and is also reflected in the building. The kindergarten classes, divided into four age groups, are arranged in the circular building in order of age from the youngest to the oldest, the latter of which proceed further to primary education. Children in each age group are taught multiple works of international literature to develop an understanding of the world around them. Until they

reach school age, the children experience several environments through the books – the forest, the village, the city, the world, and the universe. The interior and the courtyard functions reflect the themes of the literary works they explore, giving them the first-hand contact with moss, trees, meadows, lawns, yards, and sand – the things that compose the universe.

The first floor of the kindergarten houses classrooms and verandas, providing a sense of continuity between the younger and older students: the younger children's playground is visible from the first-floor corridor, and the older children's class activities are visible to the younger children. The ground floor of the school building is devoted to the public aspects of school life – the large hall, the sports area, and the children's wardrobes. The first floor, meanwhile, is dedicated to a library and specialized classrooms for languages, arts and crafts, computers, chemistry, and physics. The two wings of the structure are connected by a "dream bridge". The school wing has an entrance of its own emphasized by the large opening in the facade. The opening offers a roofed outdoor area and is intersected by a 100-meter running track for physical education classes.

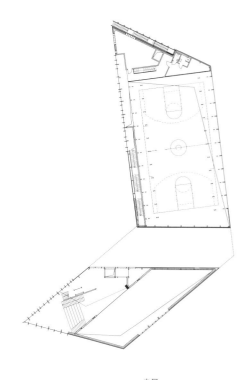

夹层
mezzanine floor

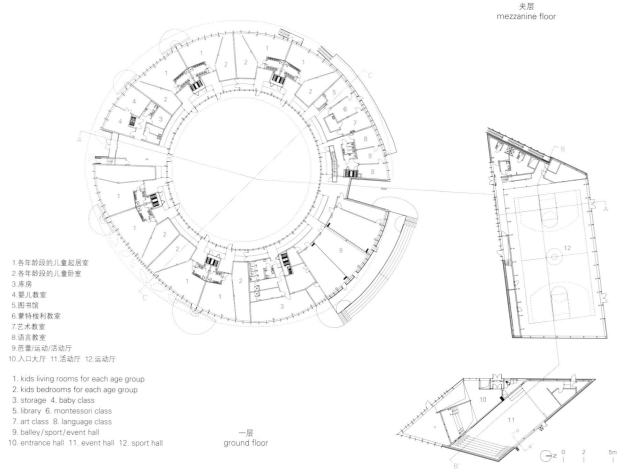

1. 各年龄段的儿童起居室
2. 各年龄段的儿童卧室
3. 库房
4. 婴儿教室
5. 图书馆
6. 蒙特梭利教室
7. 艺术教室
8. 语言教室
9. 芭蕾/运动/活动厅
10. 入口大厅 11. 活动厅 12. 运动厅

1. kids living rooms for each age group
2. kids bedrooms for each age group
3. storage 4. baby class
5. library 6. montessori class
7. art class 8. language class
9. balley/sport/event hall
10. entrance hall 11. event hall 12. sport hall

一层
ground floor

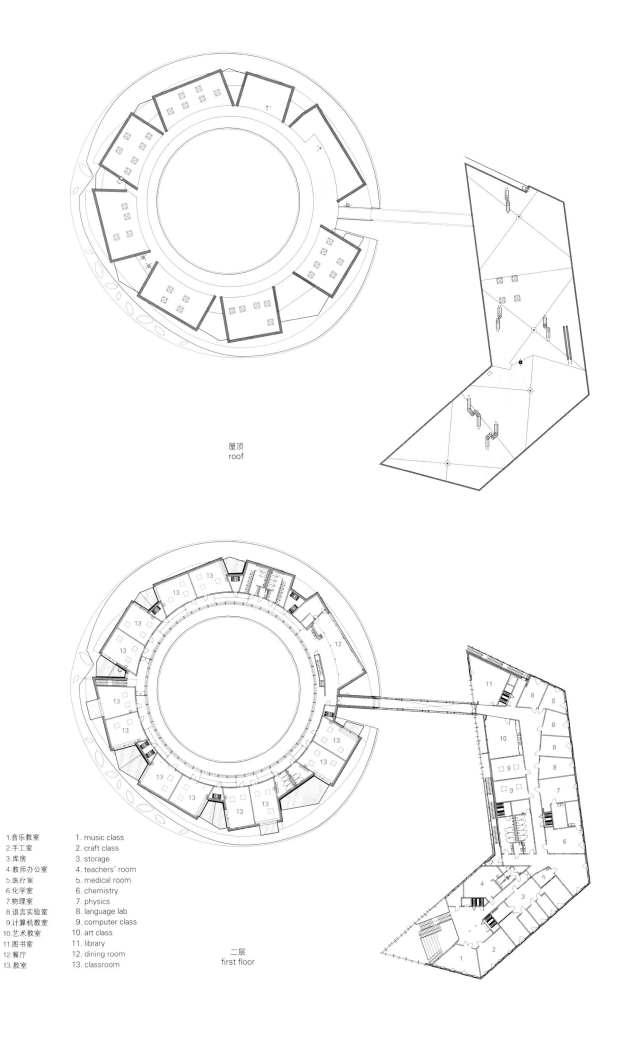

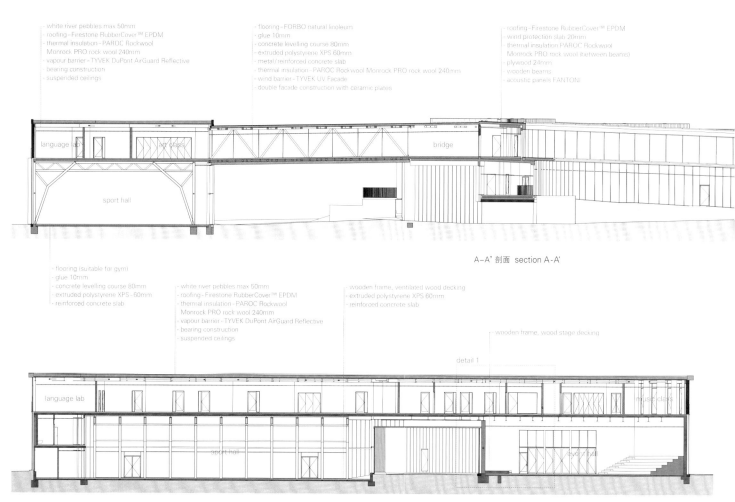

- white river pebbles max 50mm
- roofing - Firestone RubberCover™ EPDM
- thermal insulation - PAROC Rockwool Monrock PRO rock wool 240mm
- vapour barrier - TYVEK DuPont AirGuard Reflective
- bearing construction
- suspended ceilings

- flooring - FORBO natural linoleum
- glue 10mm
- concrete levelling course 80mm
- extruded polystyrene XPS 60mm
- metal/reinforced concrete slab
- thermal insulation - PAROC Rockwool Monrock PRO rock wool 240mm
- wind barrier - TYVEK UV Facade
- double facade construction with ceramic plates

- roofing - Firestone RubberCover™ EPDM
- wind protection slab 20mm
- thermal insulation PAROC Rockwool Monrock PRO rock wool (between beams)
- plywood 24mm
- wooden beams
- acoustic panels FANTONI

A-A' 剖面 section A-A'

- flooring (suitable for gym)
- glue 10mm
- concrete levelling course 80mm
- extruded polystyrene XPS 60mm
- reinforced concrete slab

- white river pebbles max 50mm
- roofing - Firestone RubberCover™ EPDM
- thermal insulation - PAROC Rockwool Monrock PRO rock wool 240mm
- vapour barrier - TYVEK DuPont AirGuard Reflective
- bearing construction
- suspended ceilings

- wooden frame, ventilated wood decking
- extruded polystyrene XPS 60mm
- reinforced concrete slab

- wooden frame, wood stage decking

detail 1

B-B' 剖面 section B-B'

The main reasons for the monolithic ferroconcrete construction were ground conditions and the spatial configuration of the building. This method has the benefits of allowing large roofs around the perimeter of the building, which follow the movement of the sun during the day, thus protecting the rooms from direct sunlight and heat.

The vertical arrangement of the aluminum facades plays a similar role, allowing direct sunlight to shine through the glass segments of the facade for short periods of time.

The large windows in the first-floor classrooms are oriented toward the northeast. The classrooms are also installed with skylights for natural light during the daytime.

The facade of the school building boasts high, narrow windows that allow more sunlight and warmth into the interior while evoking the imagery of partially-closed blinds.

The materials that compose the facade have a long lifespan and will not need renovation for some time. The wood decoration on the facades is made of Siberian larch, a perfect fit for both the local climate and sustainability purpose.

The building operates on a non-central heating, cooling, and airing system, which is monitored by a building management system. The system guarantees high-quality air circulation in any room in the kindergarten and school areas, responding to user control and exterior weather conditions. Vents incorporate microperforated textiles that provide slow, continuous air circulation while also being easy to maintain and clean. Sound-absorbing panels have been installed as suspended ceilings or wall decorations in noisy rooms (such as kindergarten groups and classrooms) to increase comfort for the users.

Plywood made of birch – a wood that reflects the mentality of the locals – is the material of choice for the interior. Natural linoleum and chalk paint with decorative print are also selected for their suitability in the children's educational development, health, and the eco-friendliness of the rooms where they will spend a majority of time for the next several years.

C-C' 剖面 section C-C'

项目名称：Exupéry International School / 地点：Jauna street 8, Pinki, Babites parish, Latvia / 建筑设计：Juris Lasis, Laura Pelse - 8 A.M. 项目团队：Elina Candere, Anastasija Pimenova, Martins Valters, Frenks Marsans, Deniss Maruhlenko, Viktorija Jakovleva / 项目管理：Ivars Nelke / 结构工程：Kurbads, Efiko, NCS LV / 道路方案：Vents Radzins / 景观：Labie koki / 声学：Arturs Perkons / 承包商：Merks / 场地面积：20,000m² / 建筑面积：6,670m² / 总楼面面积：8,000m² / 材料：SCHUCO FV 50 + GUARDIAN "SunGuard" High Performance, innovative solar control HP silver; LAMBERTS, Linit, U-Glass, Solar P26/60/7 + Wacotech; Siberian larch ("Anzage"); Birch plywood ("Latvijas finieris") 完工时间：2016 / 摄影：©Indrikis Sturmanis (courtesy of the architect)

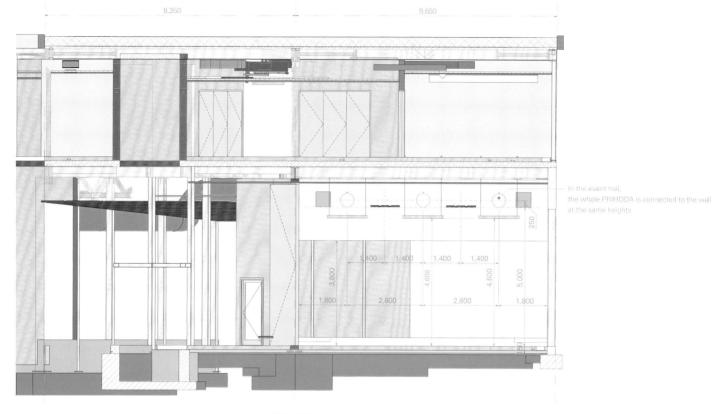

详图1_活动厅
detail 1_event hall

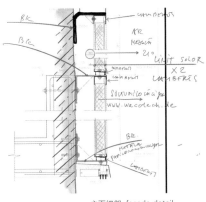
立面细部 facade detail

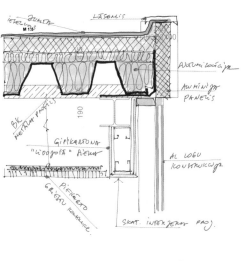

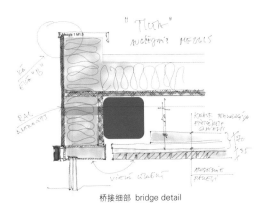
桥接细部 bridge detail

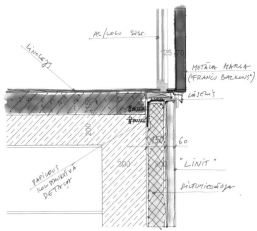

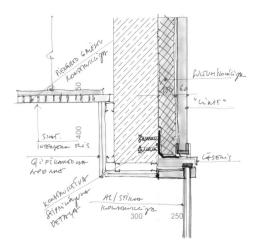

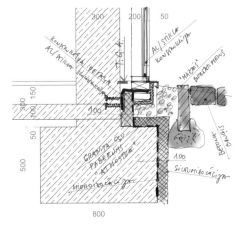
a-a' 详图 detail a-a'

玛利娅·蒙特梭利马萨特兰学校
María Montessori Mazatlán School

EPArquitectos + Estudio Macías Peredo

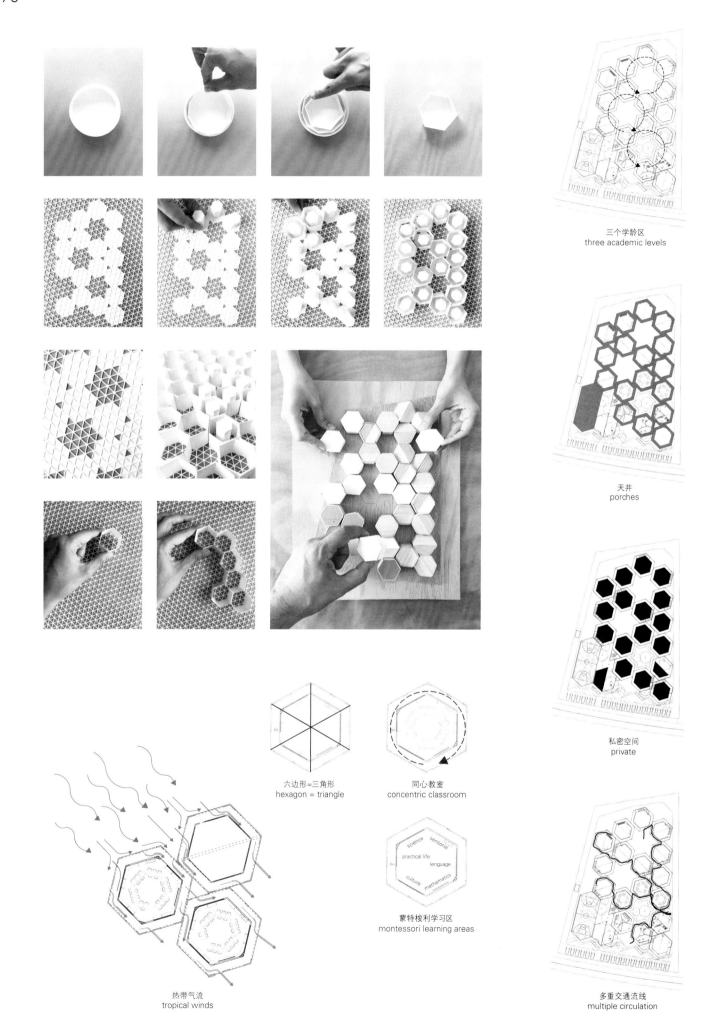

马萨特兰地处太平洋沿岸,该地气候几乎终年炎热潮湿。这就要求建筑设计不仅要考虑并解决气候条件的局限性,还要解决空气中盐度过高的问题。所提出的设计策略应在确保不折损自然光照,能与外部相通的前提下,尽量减小高温对教室的影响,并采用当地的低腐蚀性施工方法。

为控制人员的出入并保证学生的安全,该校外部只有一个外立面对城市开放,内部则采用了可控式景观设计。蒙特梭利教育模式并非传统教育体系,所以教室的设计应保持动感,能让孩子们在其中进行体验,并发展他们的感官知觉。为使教室的形式有利于动感的维持,教室内的空间均采用离心式排列方法,而非线性排列。

19个六边形模块自成体系,模块的中间为教室,四周为带有门廊的走廊,以利隔热和空气增压。另外,这种配置还解决了流线和半开放式活动的视觉通透性问题。模块之间互相连接且连接方式不同,形成了多处多面体天井,其上天窗的设计有利于自然采光及空气流通。一些高度各异的小型建筑围绕三个天井组成了一小片景观,形成了一个"儿童村"。模块内部可用于多种用途,无论是行政管理还是教学和娱乐活动都可以进行,为小至蹒跚学步、大至12岁以上的儿童们提供服务。

项目的挑战有两个层面。首先,2100m²的空心砖建筑必须在最多四个半月内竣工,所以项目十分强调这些独立模块的灵活性。其次,这些大小形状不一的三角形开口对(传统的)开洞理念提出挑战,让人们难免揣测三角形到底是不是一个适合开窗的灵活且富有趣味性的几何图形。不过,对于那些进入到"村庄"里的人来说,他们会立刻就认同建筑这种无论是对孩子还是对成人都做出回应的形态。每间教室,不论是作为一个独立的单位,还是作为整个系统的一部分,都能使孩子们自由地建立起自己的秩序,并享受其中。

Mazatlán is a city on the Pacific coast with a hot and humid climate mostly throughout the year. This invites to design an architecture that considers and solves the climate condition as well as a high salinity level in the air. The proposed strategy should minimize the heat impact on the classrooms without losing natural light and relation to the outside with local low-corrosion construction methods.
To control access and guarantee children security, the project, externally, has only one facade facing the city, and internally, makes the controlled landscape. The Montessori model is not a conventional education system, so classrooms should facilitate dynamics where children can experi-

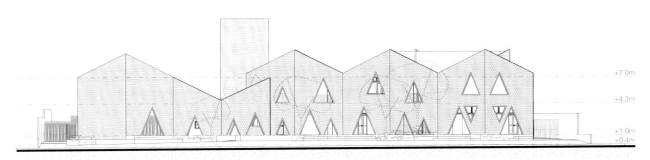

东南立面 south-east elevation

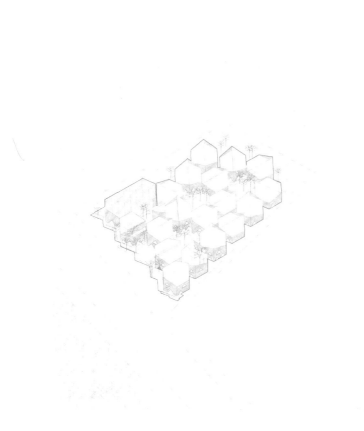

屋顶 roof

1.儿童社区
2.巢
3.儿童室
4.工作坊
5.班主任室
6.浴室
7.天井
8.行政办公室
9.厨房
10.商业

1. children's community
2. nest
3. children's room
4. workshop
5. coordination
6. bathrooms
7. porch
8. administration room
9. kitchen
10. commercial

1.儿童公社夹层
2.儿童之家夹层
3.工作坊
4.工具室
5.行政办公室

1. mezzanine of children's community
2. mezzanine of children's house
3. workshop
4. utility room
5. administration room

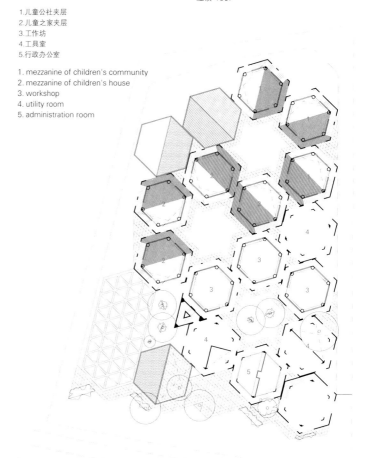

一层 ground floor

二层 first floor

项目名称：Colegio María Montessori Mazatlán
地点：Paseo del Atlántico 6208, Marina Mazatlán, Mazatlán, Sinaloa, México
建筑设计：Erick Pérez Páez - EPArquitectos;
Salvador Macías Corona, Magui Peredo Arenas - Estudio Macías Peredo
合作者：Isaac Veloz Naranjo, Guillermo Barrera Romero,
Sacnité Flores Fernández, Alejandra Garate Delgado
施工：EPArquitectos + H Arquitectos
结构工程师：Ing. Juan Jesús Aguirre Herrera - CEROMOTION
总楼面面积：2,100m²
材料：Pacific Cast Concrete - floors and lintels; Novaceramic - Tabique Novablock; PVA - glass and aluminum; FORMOTICA - wood
完工时间：2016
摄影：©Onnis Luque (courtesy of the architect)

ment and foster their senses. To achieve a form in favor of these dynamics, the classroom space is arranged centrifugally instead of linearly.

A system of 19 hexagonal modules places the classrooms inwards, creating a perimeter as the porched hallway that promotes thermal insulation and air pressurization. In turn, this configuration solves circulation and visual transparency for semi-open activities. The modules connect and shift to generate polyhedral patios that capture natural light

and air currents through skylights. Around three patios, a "children's village", a small landscape of tiny villas at different heights, comes into the world. The interior of the modules is designed to host all kind of activities, from administration and management to those teaching and recreational ones, serving younger users that range from toddlers to children older than 12 years.

The challenge of the project came in two stages. First, the 2,100m² building of hollow bricks had to be built in four and a half months at most, so the flexibility of independent modules was highly emphasized. Second, the triangular openings in varying proportions questioned the idea of an opening, arousing doubts if the triangle is an agile and playful geometry to build a window in. But those who enter the "village" spontaneously agree with the morphology that responds to the users, from kids to adults. In the classroom as a single cell and that expands to the whole system, children enjoy the freedom to build their own order.

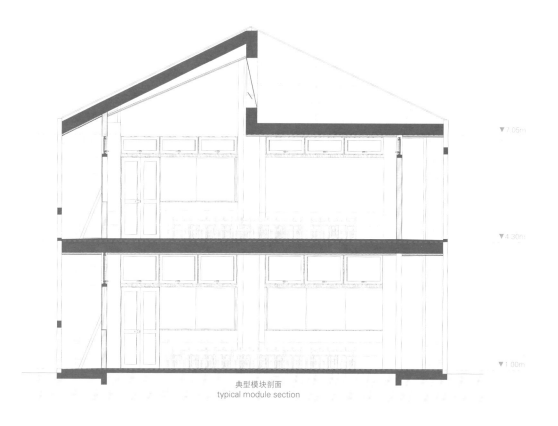

典型模块剖面
typical module section

儿童村
Children's Village

Aleph Zero + Rosenbaum

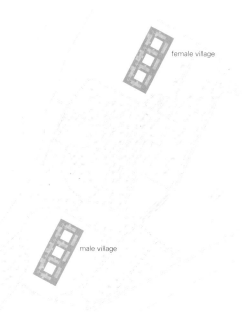

可容纳540名孩子在校学习的Canuanã寄宿学校项目旨在进行革新及文化援助,鼓励当地的施工技术,传承当地文化并保护当地的自然风光,同时帮助孩子们构建自我并找到归属感,这些都是儿童发展过程中不可或缺的因素。

随着孩子们的不断长大,他们的观点也一直在变化。为使人们更好地了解这所寄宿学校的规模,更好地推动对学习兼居住空间的理解,学校采用了一种名为"温和转换"的方法。为此,建筑师与当地社区、教师、管理部门,特别是学生展开了开放且密切的合作。整个过程中,建筑师就工作坊和动态性问题进行了研究,深入思考并和相关人员通力合作,通过现代技术和丰富的地方乡土知识间的不断交流,最终对这一问题达成了共识,并找到了可能的解决方案。

通过建筑技术的不断提高和与教师、学生们的交流,建筑师得出了一套解决方案,这个方案被认为是对场地在更大范围内进行空间组织的第一步。新布局将建造两个尺度更大(容纳人数也更多)、通风效果更佳的"村落",并按要求遵循之前就存在的类型划分设定,分别供男生和女生使用。两个"村落"并没有位于中心轴线的位置,因为中心位置的空间只提供给与学习行为直接相关的活动,而是被建在尽量靠近边缘的关键位置,这有利于起到组织整片土地空间作用的农场综合设施的发展,也有利于把学校视为一个整体而对其空间性和功能性进行解读。

在新设计中,住宿区不再由大型宿舍空间构成,而是由45个6人间寝室组成,每个寝室内设有双层床铺,更多的储物空间和独立洗漱区。这45个寝室每5个为一组,围着3个大天井分布,天井中种满了当地的草原和热带植被,这些庭院不仅是人们的聚会场所,同时也起到了降温和防潮的作用。

紧邻睡眠区的楼上建造了各种互动场所,如电视屋、阅读室、阳台、放映室、吊床休息室和游戏区等。这些补充性项目经过精选后设立在学生的生活区域,不仅能提升学生的总体生活质量,增强他们与校园生活的联系,同时也为非在校生提供了补充性的学习空间。

屋顶采用薄薄的白色金属结构,由常规的5.9m×5.9m轻质木结构作为支撑,覆盖在两个"村落"及其公共区域之上并为其遮阳避雨。除了这一优点,屋架结构还在建筑的内部和外部之间形成了一个中间地带,构成了一个标示出宽阔地平线的平台,并框起了室内外植被的景色。

之所以在结构构件中选用层压胶合桉木,是因为这种材料具有多功能性、可预制性和可持续性,以回应加快施工进度并尽量减少学校运作麻烦的需求。同理,采用由当地土壤制成的砖块进行施工,既避免了长距离的交通运输,也利用了其优质的保温性能。砖块被用来制作砖围墙及其上的格子砖图案,这样既有利于通风,也有利于保护洗漱区的安全。砖块的使用兼具技术性和美感,这种方式在当地经久不衰。

最后,新"村落"的目的在于通过设计与当地知识和建设潜力之间的对话,增强孩子们的自尊和独立意识,增强他们的归属感和保护环境的责任感,提高他们的综合学业表现。就这样,本土技艺和积极的可持续住宅模式间的对话机制形成了。

项目名称：Bradesco Foundation - Canuanã Students' Homes
地点：Formoso do Araguaia, Tocantins, Brazil
建筑设计：Aleph Zero + Rosenbaum
木结构设计、制造和施工：Ita Construtora
照明工程：Lux Lighting Projects
地基工程：Meirelles Carvalho
隔热舒适性咨询：Environmental Consulting
机电工程：Lutie
混凝土板：Trima / 承建商：Inova TS
室内家具设计：Rosenbaum and Fetiche Design
景观设计师：Raul Pereira Associated Architects
资料记录：Fabiana Zanin
声学和保温咨询：Ambiental Consultoria
业主：Bradesco Foundation - Canuanã School
总楼面面积：23,344.17m² / 施工时间：2014—2016
摄影：©Leonardo Finotti

The project to house 540 children who study at Canuanã boarding school aims towards transformation, cultural rescue, encouragement of local constructive techniques, native knowledge and beauty, together with the construction of a notion of self and belonging necessary for the children's development.

To better understand the scale of this school inhabited by children whose point of view changes with every grown inch, and to propose a new understanding of the learning space also as a place of residence, "A Gente Transforma" methodology was applied. It involved an open and intense collaboration with the local community, teachers,

administration, and especially, the children. The process went through stages of research, immersion and collaboration with all those involved in workshops and dynamics, in which a common understanding of the problem and its possible solutions arising from the dialogue between the contemporary technique and the rich local vernacular knowledge, were sought.

The process of continuous architectonic enhancement and interchange with teachers and students led to a solution imagined as the first step on the broader organization of the site. The new configuration foresees two larger in size (also in number of residents) and airier villages, one for male students and one for female students, following the required pre-existing genre divided setting. Each one is placed at a strategic point no longer set inside the central axis, which is to be filled solely with programs directly related to the act of learning, but rather closer to the edges to guide the growth of the farm complex organizing the territory and thus enabling a better spatial and functional reading of the school as a whole.

In this new moment, the residences will no longer be composed of large dormitory spaces, but rather of 45 units for six students each, with bunk beds, more storage space and private washing areas. The units are then set together

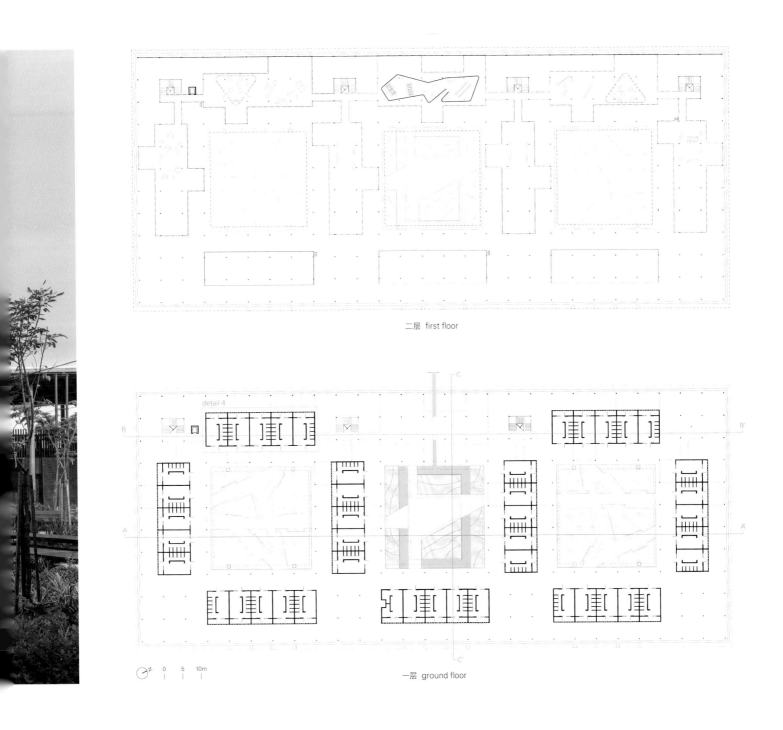

二层 first floor

一层 ground floor

in groups of five, organized around three large patios filled with local savannah and tropical species. These courtyards serve as gathering spaces and also as means of reducing heat and controlling air humidity.

Adjacent and on top of the sleeping blocks are different interacting areas such as TV rooms, reading spaces, balconies, resting hammock spaces, playing places, among many more. These complementary programs were set and refined together with the students in order to improve their overall quality of life and increase their bond to the school, and to serve as complementary learning spaces for the population outside of the school year.

A thin, white metallic roof supported by a lightweight wood structure following a regular grid of 5.9m by 5.9m embraces the villages and common spaces. Beyond protection from sun and rain, the roof and grid set composes an intermediary space between the outside and the inside, behaving as a great veranda that marks the vast horizon and frames views of the exterior and interior vegetation.

The choice for glued laminated eucalyptus wood in the

structural elements came from its versatility, pre-fabrication and sustainable characteristics, in response to the necessity of accelerating construction speed and minimizing hassle over school's functioning. Likewise, stabilized earth blocks composed of local soil were chosen for construction to dismiss long distance transportation and to utilize optimal thermal properties. The material was used as enclosure in form of apparent brick walls as well as brick latticework to provide ventilation and protection to the washing areas. The brick performed technically but also aesthetically, much in the way the locals have been doing for a long time. Ultimately, the design for the new villages aims to increase the children's self-esteem, individuality, sense of belonging, responsibility for the environment and overall academic performance through the dialogue with local knowledge and constructive potential. Thus, a dialogue is created between vernacular techniques and a positive model for sustainable housing.

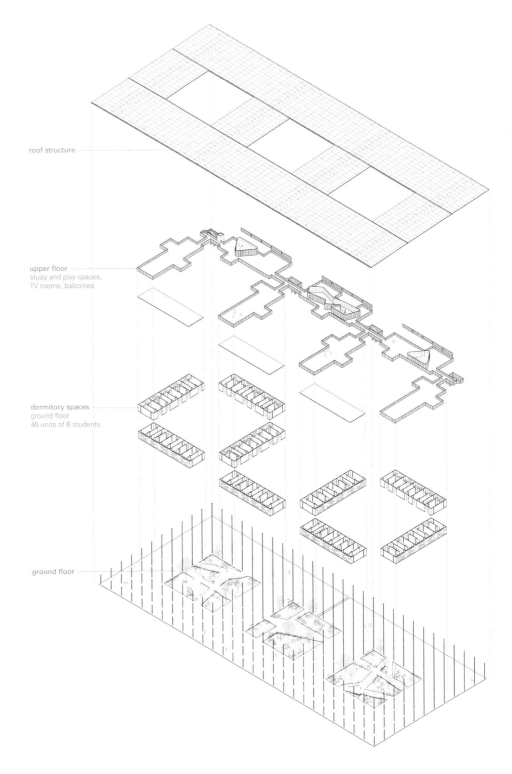

详图1 detail 1 详图2 detail 2 详图3 detail 3

屋顶_木柱梁之间的连接
roof _ wood posts and beams connection

二层_木柱梁之间的连接
first floor _ wood posts and beams connection

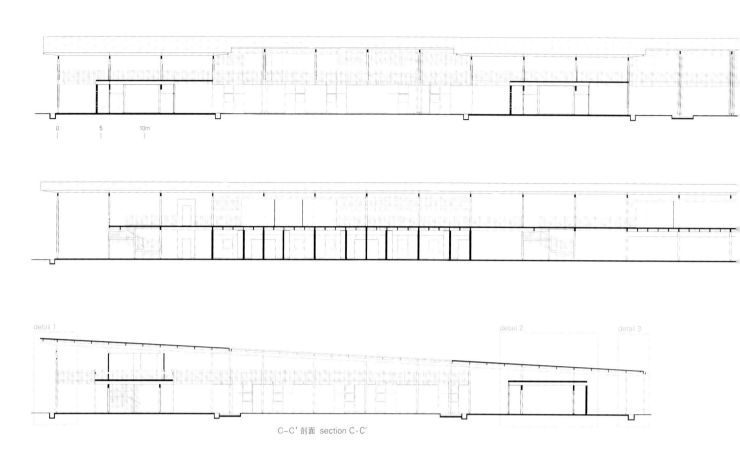

C-C' 剖面 section C-C'

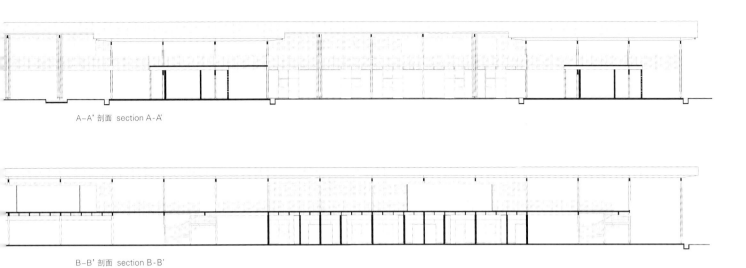

A-A' 剖面 section A-A'

B-B' 剖面 section B-B'

宿舍
Dormitory

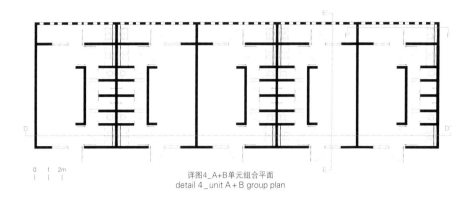

详图4_A+B单元组合平面
detail 4 _ unit A + B group plan

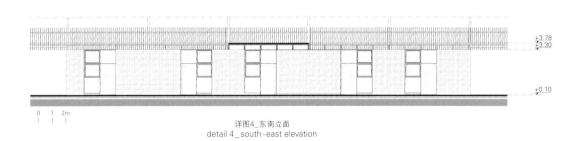

详图4_东南立面
detail 4 _ south-east elevation

详图4_西北立面
detail 4 _ north-west elevation

详图4_砖墙
detail 4 _ brick wall

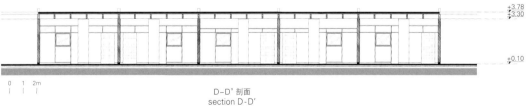

D-D' 剖面
section D-D'

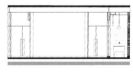

E-E' 剖面   F-F' 剖面
section E-E'   section F-F'

砖块类型
brick typologies

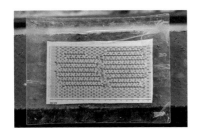

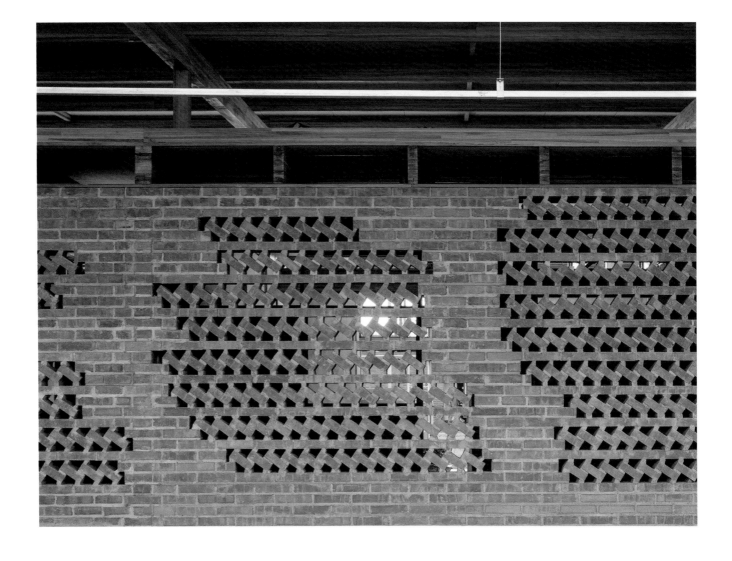

be – MINE 游乐场
Play Landscape be - MINE

OMGEVING + Carve

2015年1月，由OMGEVING+Carve（建筑师）和Krinkels（承建商）组成的团队在国际比赛中胜出，将在比利时贝林根的一处矿渣堆（或称"terril"）上建造一个富有冒险性的地标性游乐场。对矿渣山的改造是这个名为be-MINE旅游娱乐项目的一部分。贝林根是佛兰德斯地区最大的工业遗址，项目旨在为当地的标志性煤矿注入新的活力。这座曾经的矿业城市要求为60m高的碎石山赋予新的功能，并将古老的工业建筑改造为新的文化热点区，让人们以一种娱乐性的方式感受当地的历史。

该设计一共有三个部分，分别是作为标志性景物的木桩森林、山体侧面一处极富冒险性并有着凹凸棱格的游戏场以及碎石山顶的煤矿广场。它们连接了碎石山体与过往的历史。一条楼梯像脊骨一样贯通上下，串联起位于不同高度的各个空间。夜晚，沿着楼梯而设的灯带亮起，让这里起伏的地形清晰可见。

在木桩森林中感受矿业史

木桩森林重塑并突出了山体的地势起伏。1600个木桩固定排列在山体的北侧，从底部直到顶部，覆盖了整个碎石山体。这些圆柱形木桩曾被用于支撑长达数公里的地下矿井，不禁让人回想起这里的矿业史。这是一种强烈的空间形态和空间干预策略，展现出山体的规模和此处的工业传统。

木桩森林中的一部分区域专门用于具有冒险性的娱乐方式，其中设有平衡木、攀爬网、吊床、迷宫和绳索。这些木桩呈网格状分布，形成了一种有趣的视觉效果，能使人回想起过去那黑暗的矿井。

棱格游戏场

木桩森林中镶嵌着一处宽阔的棱格游戏场，场地表面随碎石山褶皱起伏，其形态从远处亦清晰可见。棱格场地极具挑战性：在山脚处，

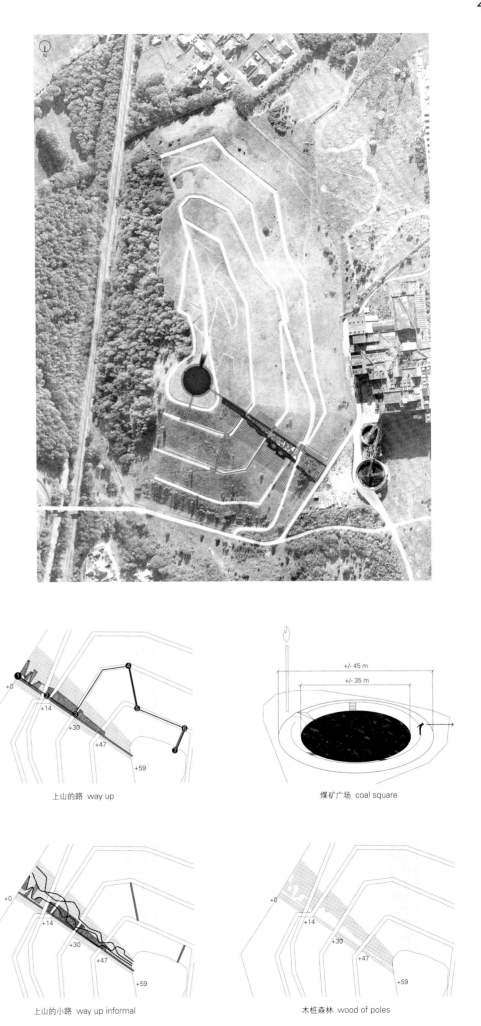

上山的路 way up

煤矿广场 coal square

上山的小路 way up informal

木桩森林 wood of poles

这些棱格犹如碎片般散落；越往高处，它却愈发狭窄。起伏的地形与穿插其中的爬行隧道、带扶手的攀援表面和巨型楼梯等娱乐项目提供了多样化的游戏方式。其中最引人注目的当属半山腰上20多米长的巨型滑梯，它随混凝土表面的地形起伏并与之融为一体。

煤矿广场

60m高的碎石山山顶处打造了一个煤矿广场，能够反映出碎石山过去与现在的特点，也见证了"黑金"煤矿的存在。煤矿广场倾斜的边缘上刻有此地以及周边矿区景观的历史信息，人们还可以坐在这里稍事休息。游客还可以在高处的岩石堆上漫步，并俯瞰Limburg矿区周围的壮丽景观。

The team of Omgeving + Carve (design) and Krinkels (contractor) won the international competition to design an adventurous play-scape and landmark on a "terril" in Beringen, Belgium, in January 2015. The adventure mountain is part of the touristic, recreative project be-MINE that aims to bring new life to the monumental coal-mining site in Beringen, the largest industrial-archeological site in Flanders. The former mining city asked to add new functions to the 60-meter high rubble mountain, and to redevelop the old industrial buildings into a cultural hotspot where its history can be experienced in a playful way.

The design consists of three parts, which create a unity with the mountain and its past: a pole forest as a landmark, an adventurous prismatic play surface on the flank of the mountain and a coal square on the top of the "terril". The spine of the ensemble as straight stairs provides access to all levels. At night, a light line along the stairs makes the topography of the terril visible.

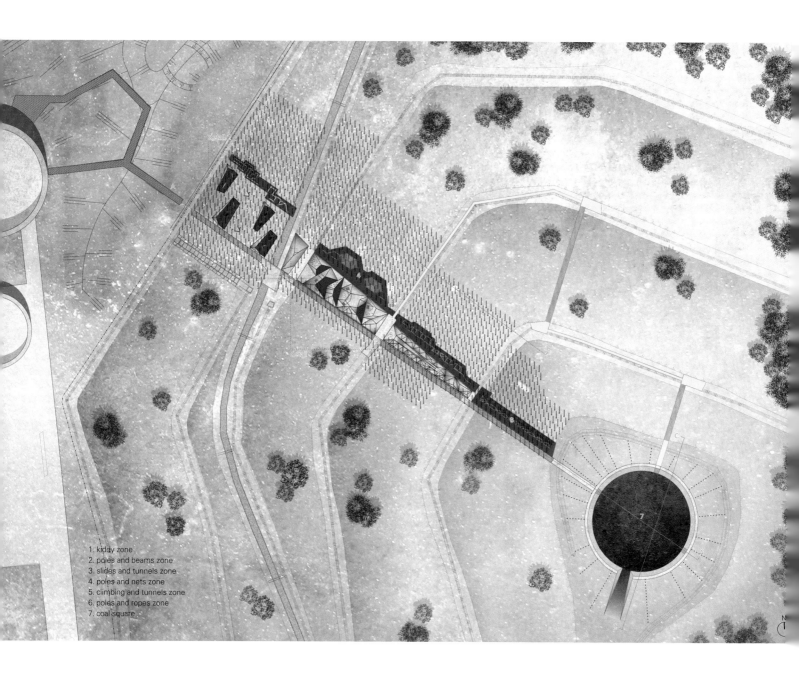

1. kiddy zone
2. poles and beams zone
3. slides and tunnels zone
4. poles and nets zone
5. climbing and tunnels zone
6. poles and ropes zone
7. coal square

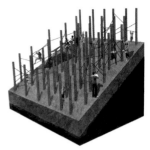
木桩和绳索区_4层
poles and ropes zone _ level 4 (47~60m)

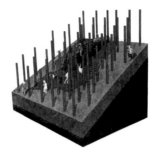
木桩和网区_2层
poles and nets zone _ level 2 (14~30m)

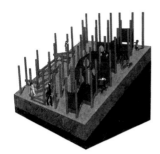
木桩和平衡木区_1层
poles and beams zone _ level 1 (0~10m)

木桩和绳索区_3层
poles and ropes zone _ level 3 (37~47m)

攀爬和隧道区_3层
climbing and tunnels zone _ level 3 (37~47m)

滑梯和隧道区_2层
slides and tunnels zone _ level 2 (14~30m)

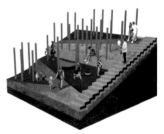
低龄区_1层
kiddy zone _ level 1 (0~10m)

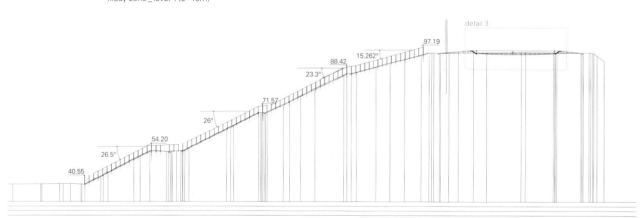
A-A' 剖面 section A-A'

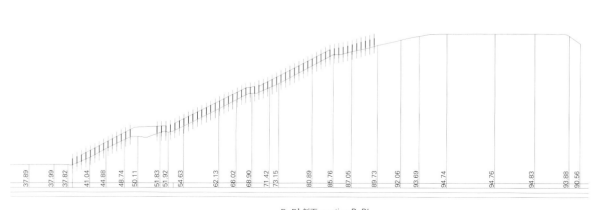
B-B' 剖面 section B-B'

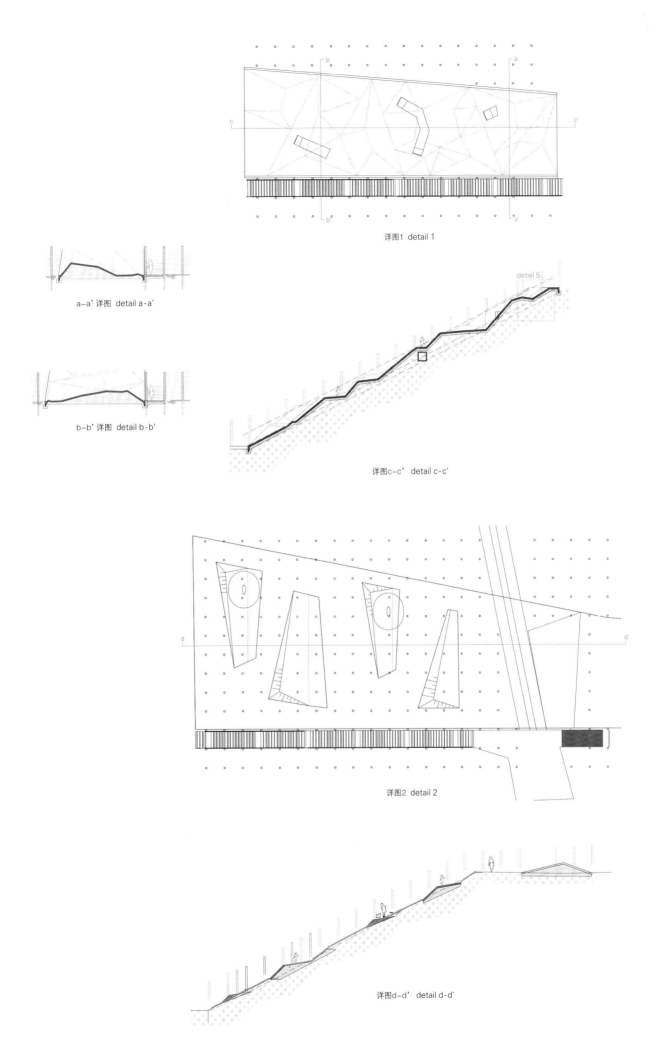

详图1 detail 1

a-a' 详图 detail a-a'

b-b' 详图 detail b-b'

详图c-c' detail c-c'

详图2 detail 2

详图d-d' detail d-d'

项目名称：Play landscape be-MINE / 地点：Beringen, Belgium / 建筑设计：OMGEVING, Carve / 项目团队：Luc Wallays, Maarten Moers, Peter Seynaeve, Peter Swyngedauw, Ada Barbu, Tom Beyaert-OMGEVING; Elger Blitz, Mark van der Eng, Jasper van de Schaaf, Hannah Schubert, Johannes Müller, Clément Gay-Carve (engineer) / 主承包商：Krinkels / 分包商：Van Vliet, IJreka / 规模：10,060m² (of which 5,200m² rope forest, 1,200m² prismatic play surface, 1,200m² coal square) / 完工时间：2016.9 / 摄影：©Benoit Meeus (courtesy of the architect)-p.206~207, p.211, p.213, p.216lower, p.218, p.219; ©Marleen Beek (courtesy of the architect)-p.212upper, p.217; ©Hannah Schubert (courtesy of the architect)-p.208, p.212lower, p.216upper

Pole Forest as Reference to the Mining Past

The topography of the landscape has regained structure and is made visible by a pole forest: 1,600 timber poles are anchored in the northern flank of the terril, from top to bottom. The rounded poles refer to the mining past; they were used for supporting the kilometers long underground mining shafts. It is a strong spatial gesture and an intervention that relates to the scale of the hill and the industrial heritage on this location.

A part of the poles has been dedicated to an adventurous play course with balancing beams, climbing nets, hammocks, a labyrinth and a rope course. The poles are placed in a grid, which results in an interesting perspective effect: the sightlines create an experience that reminds of the dark mining shafts of the past.

Prismatic Play Surface

Wedged in between the pole forest lies a large, prismatically shaped play surface. It has been "draped" over the terril following its heightlines – a gesture that is visible from afar. The surface is a challenging object that narrows down to the top and "crumbles" at the foot of the hill. This tectonic landscape offers space to an endless variation of play options, and is scattered with crawling tunnels, climbing surfaces with climbing grips and "giant stairs". Its spectacular highlight is the more than twenty meters long slide, which is placed halfway up the hill and is integrated within the relief of the concrete play surface.

Coal Square

On top of the terril, at 60m height, a "Coal Square" was created to reflect both the past and the present characters of the terril. The square visualizes the presence of the "black gold". The sloped edges of the Coal Square can be used for seating and contain historic information on the site and the surrounding mining landscape. Visitors can take a stroll on the raised talud and enjoy the panoramic views on the surrounding Limburg mining landscape.

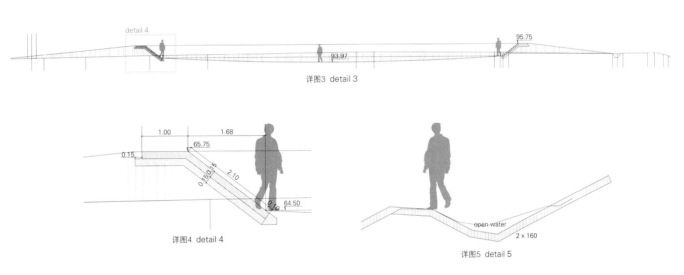

详图3 detail 3

详图4 detail 4

详图5 detail 5

C-C' 剖面 section C-C'

El Equipo Mazzanti

Giancarlo Mazzanti was born in Barranquilla, Colombia in 1963. Graduated with a degree in Architecture from the Pontifical University of Javeriana in Bogotá, Colombia in 1987. Received a postgraduate degree in history and theory of architecture and industrial design from the University of Florence, Italy in 1991. Has taught in several Colombian Universities and some of the most prestigious American Universities such as Princeton University and Harvard Graduate School of Design (GSD). Won the XX Colombian Architecture Biennial, Ibero-American Biennial, Panamerican Architecture Biennial and the Global Award for Sustainable Architecture prize from the French Institute of Architecture.

Rosenbaum

Was founded by Marcelo Rosenbaum, born in 1968, Sao Paulo. He has been operating the office for more than 20 years. Passionately interpreting Brazilian values through design. Has been the creative artist of the "Home Sweet Home" section produced by Caldeirão do Huck TV show on TV network Rede Globo for five years, and has delivered lectures for different industry sectors in his domain. Adriana Benguela is a Brazilian architect, who builds the history of Rosenbaum together with Marcelo Rosenbaum since 1997.

KLab Architecture

Konstantinos Labrinopoulos was born in 1970 in Athens, Greece. Graduated from the School of architecture, National Technical University of Athens in 1994. Continued his studies in the United States and received his Master III degree in Advanced Architectural Design from the SCI-Arc, Los Angeles in 1996. Co-founded Klmf in 2001 and founded KLab Architecture in 2007. Was a visiting professor at Bezalel Academy of Art and Design in Jerusalem, a visiting critic at the Hochschule fur Technik Stuttgart Architecture School. Currently is teaching at the Architecture School of the University of Patras. His work has been shortlisted for several awards as Mies Van Der Rohe award, Piranesi award among others. KLab Architecture has been included as one of the most upcoming architecture practices in the world by Wallpaper magazine in 2008.

Feld72

Is an Austrian architectural practice, founded in Vienna, Austria in 2002. Is led by 5 partners of different nationalities, gathered through diverse universities and ERASMUS program. They try to create positive long-term change within environments through intelligent and precise interventions. They consider their work an interface between architecture, art and applied urbanism. Anne Catherine Fleith (1975, Colmar, France) studied Architecture at EAS Strasbourg and TU Delft. Michael Obrist (1972, Bozen, Italy) studied Architecture at TU Vienna and SA Portsmouth. Richard Scheich (1972, Launceston, Australia) studied Architecture at TU Vienna and CVUT Prague. Peter Zoderer (1973, Bozen, Italy) and Mario Paintner (1973, Klagenfurt, Austria) studied Architecture at TU Vienna and TU Delft.

P106 **Yasutaka Yoshimura Architects**

Is an architecture and urban design firm based in Tokyo. Yasutaka Yoshimura was born in 1972 in Toyota, Aichi, Japan. Received his Bachelor's degree in 1995 and a Master's degree in 1999 from the Waseda University. Has worked at MVRDV, Rotterdam (1999-2001) and founded Yasutaka Yoshimura Architects Inc in 2005. Has been an adjunct lecturer of Waseda University, University of Tokyo, Tokyo Institute of Technology and others since 2002. Currently is a professor of Meiji University from 2013 and Waseda University from 2018. Received numerous prizes from the AIJ awards, WADA award, AP Award, JCD Design Award, Good Design Award, Asia Design Award and so on.

P26 **Studio3 - Institute for Experimental Studies UIBK**

The Studio3, Department of Experimental Architecture, works and teaches at the interface of contemporary art, culture and experimental architecture, aiming to encourage the emergence of a new social understanding of architecture. On the assumption that research always has something to do with discovery and that discovery is deeply affected by aspects of the new, the unknown and the future, the Institute researches the idea of "Concrete Utopias" as the innovative and visionary implementation being a basic requirement for an architecture which decisively rejects the dictates of definitions based on taste. The Playroom in University of Innsbruck was completed by 21 students under 2 supervisors, Verena Rauch and Walter Prenner.

P82 **Hibinosekkei + Youji no Shiro**

Is the name of a section of Hibino Sekkei Architecutre based in Kanagawa, Japan. The company was founded in 1972 and had launched the section which specializes in the design of spaces for children in 1991, reflecting the rapidly altering social situation. Under the circumstances of declining birth rate and the population of children, they recognized existing preschool architecture are rather inefficient and characterless. During their 24 years in operation, they have created approximately 350 spaces for children all over Japan. The spaces include newly built architecture, converted architecture, interiors and the like.

Alison Killing

Is an architect and urban designer based in Rotterdam, the Netherlands. Has written for several architecture and design magazines in the UK, contributing features and reviews to *Blueprint* and *Icon* and editing the research section of Middle East Art Design and Architecture. Most recently, she has worked as a correspondent for the online sustainability magazine *Worldchanging*. Has an eclectic design background, ranging from complex geometry and structural engineering, to humanitarian practice, to architecture and urban design. Has worked in the UK and the Netherlands, also more widely in Europe, China and Russia.

P194 **Aleph Zero**

Was co-founded by Brazilian architects, Gustavo Utrabo[left] and Pedro Duschenes[right] in 2012 and based in São Paulo. Received the awards Building of the Year 2018 of ArchDaily, APCA 2017 Award[Paulista Association of Art Critics] in the category: Architecture in Brazil. Gustavo(Curitiba, 1984) received a degree in Architecture and Urbanism from the Federal University of Paraná in Curitiba in 2010. In 2014, he also completed a specialization course in National History and Literature from UTFPR. Currently is an assistant professor of architecture at Escola da Cidade in São Paulo. Pedro(Curitiba, 1986) performed part of his studies in Munich, Germany. Graduated from the Department of Architecture and Urbanism, Federal University of Paraná in 2011.

P92　KIENTRUC O

Was co-founded by dynamic duo of principal architect, ĐÀM VU[picture-left] and industrial designer LÊ AN-NI in Sài Gòn, Vietnam. ĐÀM VU graduated from the University of Architecture, Saigon in 2001 and worked for 4 years at the T.A.D. Founded One Architecture Studio in 2007 and reformed it to KIENTRUC O in 2013. Has been teaching at the University of Architecture Ho Chi Minh City since 2015. LÊ AN-NI studied architecture at the University of Architecture, Saigon and Industrial Design at the Faculty of Engineering, LTH (Lund University). Has been teaching Industrial Design at the University of Architecture, Ho Chi Minh City since 2014.

P212　OMGEVING

Is an independent, multi-disciplinary group of architects, landscape architects, engineers, urban planners, and experts in environmental planning based in Antwerp, Belgium. From a small office of three, it evolved over the years into a large enterprise with more than 60 experts, working both for the public and private sectors. The Flemish word "OMGEVING" stands for 'surroundings' and it stresses the importance of the space around us. It includes the immediate space surrounding oneself, one's home, one's neighbourhood, region, and even country.

P52 Mario Cucinella Architects

Was founded by Mario Cucinella in Paris in 1992. Is now operating in Bologna (Italy) with an international team of more than 50 architects, engineers and designers. Sustainable building design and the rational use of energy is one of the central concerns in its work and research. MCA's response to the environment includes the collection of data on the socio-economic contexts and vernacular architecture along with the analysis of local climate and eco-systems. Strives to express, not its own architectural style, but rather the identity of places, people, materials and environments in each place. Was awarded International Fellowship from RIBA in 2016 and Honorary Fellowship Award by AIA in 2017.

Martha Thorne

Is Dean of IE School of Architecture and Design. Since 2005, she has been serving as the Executive Director of the Pritzker Architecture Prize. Prior to joining IE University, she was Associate Curator of the department of architecture at The Art Institute of Chicago. Is the co-author of the books *Masterpieces of Chicago Architecture* and *Skyscrapers: The New Millennium*, and author of numerous articles for architectural journals and encyclopedias. Received a master of city planning degree from the University of Pennsylvania and a bachelor of arts degree in urban affairs from the State University of New York at Buffalo. Currently serves on an international jury for the award ArcVision: Women and Architecture, a prize honoring outstanding women architects.

P60 PAL Design Group

Design Partner, Joey Ho was born in Taiwan, China and raised in Singapore. Received his Bachelor of Architectural Studies from The National University of Singapore and Master of Architecture from The University of Hong Kong. Established a high profile over Singapore, India, the US and Australia. Is the Chairman of the Hong Kong Interior Design Association, the director of Hong Kong Design Centre, course consultant of Hong Kong Institute of Vocational Education and advisory committee member of Hong Kong Trade and Development Council.

P180   Estudio Macías Peredo

Is led by Salvador Macías Corona and Magui Peredo Arenas, based in Guadalajara, Mexico. They have given several lectures at various universities and forums in Mexico, New York, etc., including the University of Tokyo and the University of New Mexico. Obtained the second place in the competition of the Mexico Pavilion in the Shanghai Expo 2010 and first place in the Eco Pavilion 2013 in Mexico City. Has been Awarded by The Architectural League of New York with the Emerging Voices award 2014. Exhibited their works in LIGA, Space for Architecture 2015 in Mexico City.

P18   Matter Design

Is an interdisciplinary design practice founded in 2008 by Brandon Clifford and Wes McGee. Director, Brandon Clifford received his Master of Architecture from Princeton University and his Bachelor of Science in Architecture from Georgia Tech. Senior Designer, Wes McGee received his Master of Industrial Design and his Bachelor of Mechanical Engineering from Georgia Tech. Currently Brandon serves as an assistant professor at the Massachusetts Institute of Technology and McGee is an assistant professor at the University of Michigan.

P126   Tezuka Architects

Was co-founded by Takaharu Tezuka[left] and Yui Tezuka[right] in Tokyo, Japan in 1994. Takaharu Tezuka (Tokyo, 1964) received his B. Arch. from Musashi Institute of Technology in 1987 and M. Arch. from University of Pennsylvania in 1990. Has worked for 4 years at Richard Rogers Partnership after graduation. Has been a Professor of Tokyo City University since 2009. Yui Tezuka (Kanagawa, 1969) received her B. Arch. from Musashi Institute of Technology in 1992 and continued her study at the Bartlett School of Architecture. Was a visiting faculty member at Toyo University and Tokai University. They gave lectures as Visiting Professors at the Salzburg Summer Academy and the University of California, Berkeley.

P18   FR|SCH Projects

Is an architecture and design studio founded by Michael Schanbacher and Kerri Frick. They see architecture as something greater than building. Creating spaces which connect to and enrich the lives of the people that use them. It is through a combination of spatial design, research, and an understanding of place that they define their work. Michael Schanbacher is currently an architect at Embarc Studio in Boston while Kerri Frick is the director of Intermediate Architecture Studios at the Boston Architectural College.

P180   EPArquitectos

Was established in 2000 as a collaboration platform to deal with architecture and urbanism, always taking into account the characteristics of a specific site, such as climate, culture and regional construction techniques. Emphasizes contribution to the city through their projects, constantly adding wills and passion to the development of ideas and proposals for public space to improve the quality of life. Developed art, housing, interior design, commercial, institutional and urban projects that seek to enhance the human spirit through their function, materiality, and proportion. Has been selected for the digital archive on the Mexican Pavilion at the 2016 Venice Biennale with his Carpa Olivera project in collaboration with Colectivo Urbano. Has exhibited works in different universities and forums around Mexico.

P164   8 A.M.

Is a Latvian architecture practice, co-founded in 2002 by Juris Lasis and Eduards Beernaerts. Both of them were born in Riga, Latvia and studied architecture at Riga Technical University. Laura Pelse, born in Riga, 1986, studied architecture at Riga Technical University and Technical University of Valencia. Their overarching goal is to promote well being for individuals, society, and the natural environment through sustainable architecture.

P212 **Carve**

Is a design and engineering bureau that focuses on the planning and development of public space, particularly for children and young people. Their ideas are translated into playable public spaces, parks, play elements and furniture. Carve has years of experience in designing skateparks and unconventional public spaces aiming at youth. Clients are local governments, architects, landscape architects and manufacturers of playground equipment. Since its establishment in 1997 by Elger Blitz and Mark van der Eng the office has grown into a company within which several design disciplines meet, from industrial design to civil engineering, architecture and landscape architecture.

P136 **Grupo Garoa Arquitetos Associados**

Located in São Paulo, Groupo Garoa Arquitetos Associados is a collective laboratory of architecture managed by Alexandre Gervásio, Erico Botteselli, Lucas Thomé and Pedro De Bona. Is a laboratory, because they take practice by the experimentation based on the specific constraints of each situation, not having preconceived methods of guiding. Is a collaborative, because they develop plural and concise works in which the different and complementary qualities of the team members add up. Develops projects and manages respective construction so that the drawings inform and modify the procedures, and vice versa.

© 2019 大连理工大学出版社

版权所有·侵权必究

图书在版编目(CIP)数据

儿童空间：汉英对照 / (美) 玛莎·索恩等编；司炳月，凌玥瑶，王晓华译. — 大连：大连理工大学出版社，2019.8
(建筑立场系列丛书)
ISBN 978-7-5685-2194-9

Ⅰ. ①儿… Ⅱ. ①玛… ②司… ③凌… ④王… Ⅲ. ①儿童－房间－室内装饰设计－汉、英 Ⅳ. ①TU241.049

中国版本图书馆CIP数据核字(2019)第175624号

出版发行：大连理工大学出版社
　　　　　(地址：大连市软件园路80号　邮编：116023)
印　　刷：上海锦良印刷厂有限公司
幅面尺寸：225mm×300mm
印　　张：14.25
出版时间：2019年8月第1版
印刷时间：2019年8月第1次印刷
出 版 人：金英伟
统　　筹：房　磊
责任编辑：房　磊
封面设计：王志峰
责任校对：张昕焱
书　　号：978-7-5685-2194-9
定　　价：258.00元

发　　行：0411-84708842
传　　真：0411-84701466
E-mail：12282980@qq.com
URL：http://dutp.dlut.edu.cn

本书如有印装质量问题，请与我社发行部联系更换。

建筑立场系列丛书 01：
墙体设计
ISBN: 978-7-5611-6353-5
定价：150.00元

建筑立场系列丛书 09：
墙体与外立面
ISBN: 978-7-5611-6641-3
定价：180.00元

建筑立场系列丛书 17：
旧厂房的空间蜕变
ISBN: 978-7-5611-7093-9
定价：180.00元

建筑立场系列丛书 25：
在城市中转换
ISBN: 978-7-5611-7737-2
定价：228.00元

建筑立场系列丛书 33：
本土现代化
ISBN: 978-7-5611-8380-9
定价：228.00元

建筑立场系列丛书 41：
都市与社区
ISBN: 978-7-5611-9365-5
定价：228.00元